Springer Series in Operations Research and Financial Engineering

Series editors

Thomas V. Mikosch
Sidney I. Resnick
Stephen M. Robinson

More information about this series at http://www.springer.com/series/3182

Vassili N. Kolokoltsov · Oleg A. Malafeyev

Many Agent Games in Socio-economic Systems: Corruption, Inspection, Coalition Building, Network Growth, Security

 Springer

Vassili N. Kolokoltsov
Department of Statistics
University of Warwick
Coventry, UK

Faculty of Applied Mathematics
and Control Processes
Saint Petersburg State University
Saint Petersburg, Russia

Oleg A. Malafeyev
Faculty of Applied Mathematics
and Control Processes
Saint Petersburg State University
Saint Petersburg, Russia

ISSN 1431-8598   ISSN 2197-1773   (electronic)
Springer Series in Operations Research and Financial Engineering
ISBN 978-3-030-12373-4   ISBN 978-3-030-12371-0   (eBook)
https://doi.org/10.1007/978-3-030-12371-0

Library of Congress Control Number: 2019930636

Mathematics Subject Classification (2010): 91A06, 91A13, 91A22, 91A25, 91A40, 91A80, 91B62, 91B74, 60J27, 60J28, 49J15

This Springer imprint is published by the registered company Springer Nature Switzerland AG
The registered company address is: Gewerbestrasse 11, 6330 Cham, Switzerland

Preface

The general picture of game-theoretic modeling dealt with in this book is characterized by a set of big players, also referred to as principals or major agents, acting in the background of large pools of small players, where the impact on the overall evolution of the behavior of each small player in a group decreases as the size of the group increases.

Two approaches to the analysis of such systems are clearly distinguished and are dealt with in Parts I and II.

(1) Players in groups are not independent rational optimizers. Either they are directly controlled by principals and serve the interests of the latter (pressure and collaboration setting), or they resist the actions of the principals (pressure and resistance setting) by evolving their strategies in an "evolutionary manner" via interactions with other players subject to certain clear rules, deterministic or stochastic. Such interactions, often referred to as myopic or imitating, include the exchange of opinions or experience, with some given probabilities of moving to more profitable strategies. They can also evolve via the influence of social norms.

Examples of the real-world problems involved include government representatives (often referred to in the literature as benevolent dictators) chasing corrupt bureaucrats, inspectors chasing tax avoiders, police acting against terrorist groups, and models describing attacks by computer or biological viruses. These examples include the problem of optimal allocation of a budget or of different strategies of a big player to affect small players, for instance the allocation of funds (corrected in real time) for the financial support of various business or research projects. Another class of examples concerns appropriate (or better, optimal) management of complex stochastic systems consisting of large numbers of interacting components (agents, mechanisms, vehicles, subsidiaries, species, police units, robot swarms, etc.), which may have competitive or common interests. Such management can also deal with the processes of the merging and splitting of functional units (say firms or banks) or the coalition-building of agents. The actions of the big players effectively control the distribution of small players among their possible strategies and can influence the rules of their interaction. Several big players can also compete for more effective pressure on small players. This includes, in particular, the controlled extensions of general (nonlinear) evolutionary games. Under our approach, the classical games of evolutionary biology, such as the hawk and dove game, can be recast as a controlled process of the propagation of violence, say in regions with mixed cultural and/or religious traditions.

For discrete state spaces, games of this kind were introduced in [135, 136, 138] under the name of nonlinear Markov games. Similar models with continuous state spaces were developed in [40] under the name of mean-field-type games; see also [72].

(2) Small players in groups are themselves assumed to be rational optimizers, though in the limit of a large number of players, the influence of the decisions of each individual player on the whole evolution becomes negligible. Games of this type are referred to as mean field games. They were introduced in [118] and [159], with the finite state space version put forward in [95], and since then developed into one of the most active directions of research in game theory. We shall discuss this setting mostly in combination with the first approach, with evolutionary interactions (more precisely, pressure and resistance framework) and individual decision-making taken into account simultaneously. This combination leads naturally to two-dimensional arrays of possible states of individual players, one dimension controlled by the principals and/or evolutionary interactions and the other dimension by individual decisions.

Carrying out a traditional Markov decision analysis for a large state space (large number of players and particles) is often infeasible. The general idea for our analysis is that under rather general assumptions, the limiting problem for a large number of agents can be described by a quite manageable deterministic evolution, which represents a performance of the dynamic law of large numbers (LLN). This procedure turns the "curse of dimensionality" into the "blessing of dimensionality." As we show, all basic criteria of optimal decision-making (including competitive control) can be transferred from the models of a large number of players to a simpler limiting decision problem.

Since even the deterministic limit of the combined rational decision-making processes and evolutionary-type models can become extremely complex, another key idea of our analysis is in searching for certain reasonable asymptotic regimes in which explicit calculations can be performed. Several such regimes are identified and effectively used.

We will deal mostly with discrete models, thus avoiding technicalities arising in general models (requiring stochastic differential equations or infinite-dimensional analysis). Extensions dealing with general jump-type Markov processes are often straightforward; see, e.g., [140].

From the practical point of view, the approaches developed here are mostly appropriate for dealing with socioeconomic processes that are not too far from an equilibrium. For such processes, the equilibria play the role of the so-called turnpikes, that is, attracting stationary developments. Therefore, much attention in our research is given to equilibria (stationary points of the dynamics) and their structural and dynamic stability. For processes far from equilibria, other approaches seem to be more relevant, for instance the methods for the analysis of turbulence.

Another problem needed to be addressed for concrete applications of our models lies in the necessity to get hold of the basic parameters entering its formulation, which may not be that easy. Nevertheless, the strong point of our approach is that it requires one to identify just a few real numbers, which may be derived in principle from statistical experiments or field observations, and not any unknown multidimensional distributions.

In the first, introductory, chapter we explain the main results and applications with a minimal use of technical language, in the hope of making the key ideas accessible to readers with only rudimentary mathematical background. The core of the book is intended for readers with some basic knowledge of probability, analysis, and ordinary differential equations.

For the sake of transparency we systematically develop the theory with a sequential increase in complexity, formulating and proving results first in their simplest versions and then extending generality.

Coventry, UK/Saint Petersburg, Russia   Vassili N. Kolokoltsov
Saint Petersburg, Russia   Oleg A. Malafeyev

Acknowledgements

The authors are grateful to many colleagues for fruitful discussions and especially to our collaborators on related research developments: Yu. Averbukh, A. Bensoussan, A. Hilbert, S. Katsikas, L. A. Petrosyan, M. Troeva, W. Yang. We also thank R. Avenhaus, N. Korgyn, V. Mazalov, and the anonymous referees for a careful reading of preliminary versions and making important comments.

The authors gratefully acknowledge support of RFBR grants 17-01-00069 (Part II) and 18-01-00796 (Part I).

Contents

Abbreviations

l.h.s.	Left-hand side
LLN	Law of large numbers
ODE	Ordinary differential equation
PDE	Partial differential equation
PDO	Partial differential operator
r.h.s.	Right-hand side

Basic Notations

The notation introduced here is used in the main text systematically without further comment: $\mathbf{Z}$, $\mathbf{C}$, $\mathbf{R}$, $\mathbf{N}$ denote the sets of integers, complex numbers, real numbers, and natural numbers, while $\mathbf{Z}_+$ and $\mathbf{R}_+$ are the subsets of nonnegative elements of the corresponding sets.

The letters $\mathbf{E}$, $\mathbf{P}$ are reserved for the expectation and probability with respect to various Markov chains.

For a convex closed subset Z of $\mathbf{R}^d$, let $C(Z)$ denote the space of bounded continuous functions equipped with the sup-norm: $\|f\| = \|f\|_{\text{sup}} = \sup_x |f(x)|$, and $C_\infty(Z)$ its closed subspace of functions vanishing at infinity.

The vectors in $x \in \mathbf{R}^d$ can be looked at as functions on $\{1, \ldots, d\}$, in which case the natural norm is the sup-norm $\|x\|_{\text{sup}} = max_i |x_i|$ (rather than more standard rotation-invariant Euclidean norm). But much often we shall interpret these vectors as measures on $\{1, \ldots, d\}$, in which case the natural norm becomes the l_1-norm or the integral norm $\|x\| = \|x\|_1 = \sum_j |x_j|$. This norm will be used as default in $\mathbf{R}^d$, unless stated otherwise in specific cases. Accordingly, for functions on $Z \subset \mathbf{R}^n$ we define the Lipschitz constant

$$\|f\|_{Lip} = \sup_{x \neq y} \frac{|f(x) - f(y)|}{\|x - y\|}. \tag{1}$$

As is easy to see, this can be equivalently rewritten as

$$\|f\|_{Lip} = \sup_j \sup \frac{|f(x) - f(y)|}{|x_j - y_j|}, \tag{2}$$

where the last sup is the supremum over the pairs x, y that differ only in its jth coordinate. By $C_{bLip}(Z)$ we denote the space of bounded Lipschitz functions with the norm

$$\|f\|_{bLip} = \|f\| + \|f\|_{Lip}. \tag{3}$$

Norms of $d \times d$ matrices are identified with their norms as linear operators in $\mathbf{R}^d$: $\|A\| = \sup_{x \neq 0} \|Ax\|/\|x\|$. This depends on the norm in $\mathbf{R}^d$. We will look at matrices as acting by usual left multiplications $x \to Ax$ in $\mathbf{R}^d$ equipped with the sup-norm or as acting by the right multiplications $x \to xA = A^T x$ in $\mathbf{R}^d$ equipped with the integral norm, so that in both cases (as is checked directly),

$$\|A\| = \sup_i \sum_j |A_{ij}|. \tag{4}$$

By $C^k(Z)$ we denote the space of k times continuously differentiable functions on Z with uniformly bounded derivatives equipped with the norm

$$\|f\|_{C^k(Z)} = \|f\| + \sum_{j=1}^{k} \|f^{(j)}(Z)\|,$$

where $\|f^{(j)}(Z)\|$ is the supremum of the magnitudes of all partial derivatives of f of order j on Z. In particular, for a differentiable function, $\|f\|_{C^1} = \|f\|_{bLip}$ (by (2)). For a function $f(t, x)$ that may depend on other variables, say t, we shall also use the notation

$$\|f^{(j)}\| = \left\|\frac{\partial^j f}{\partial x^j}\right\| = \left\|f(t, .)^{(j)}\right\|,$$

if we need to stress the variable x with respect to which the smoothness is assessed. We shall also use the shorter notation $\|f^{(j)}\|$, C^k or C_{bLip} for the norms $\|f^{(j)}(Z)\|$ and the spaces $C^k(Z)$ or $C_{bLip}(Z)$ when it is clear which Z is used.

The space $C(Z, \mathbf{R}^n)$ of bounded continuous vector-valued functions $f = (f_i) : Z \to \mathbf{R}^n$, $Z \subset \mathbf{R}^d$, will be also usually denoted $C(Z)$ (with some obvious abuse of notation), their norm being

$$\|f\| = \sup_{x \in Z} \|f(x)\| = \sup_{x \in Z} \sum_i |f_i(x)|, \quad \|f\|_{\sup} = \sup_{i,x} |f_i(x)|.$$

Similar shorter notation will be used for spaces of smooth vector-valued functions $f = (f_i) : Z \to \mathbf{R}^n$ equipped with the norm

$$\|f\|_{C^k(Z)} = \|f\| + \sum_{j=1}^{k} \|f^{(j)}\|, \quad \|f^{(j)}\| = \|f^{(j)}(Z)\| = \sum_{i} \left\|f_i^{(j)}(Z)\right\|.$$

In Chapter 4 we will work with the functions on infinite sequences, namely on the space l_1 of sequences $\{x = (x_1, x_2, \ldots)\}$ having finite norm $\|x\| = \sum_j |x_j|$. All the notations above will be used then for closed convex subsets Z of l_1.

Chapter 1
Introduction: Main Models and LLN Methodology

The first sections of this chapter are devoted to a brief introduction to our basic model of controlled mean-field interacting particle systems, its principal features in the framework of pressure and resistance, and to the fundamental paradigm of the scaling limit (dynamic law of large numbers), allowing one to reduce the analysis of these Markov chains with exponentially large state spaces to a low-dimensional dynamic system expressed in terms of simple ordinary differential equations (ODEs), so called kinetic equations. The rigorous mathematical justification of this reduction is postponed to the next chapter, while in the rest of this chapter we describe various classes of real-life socioeconomic processes fitting the general scheme: inspection, corruption, cybersecurity, counterterrorism, coalition-building, merging and splitting, threshold control behavior in active network structures, evolutionary games with controlled mean-field dependent payoffs, and many others. In fact, our general framework allows us to bring a great variety of models (often studied independently) under a single umbrella that is amenable to a simple unified analysis. Moreover, many standard socioeconomic models dealt with traditionally via small games with two or three players arise in our presentation in new clothing, allowing for the analysis of many-player extensions. In the last section we give a brief guide to the further content of the book.

1.1 What Is a Markov Chain

The notion of Markov chains, which provide the simplest models of random evolutions, is a cornerstone of our analysis. Here we describe the basic ideas with a minimal use of technical terms.

One distinguishes Markov chains in discrete and continuous time.

© Springer Nature Switzerland AG 2019
V. N. Kolokoltsov and O. A. Malafeyev, *Many Agent Games in Socio-economic Systems: Corruption, Inspection, Coalition Building, Network Growth, Security*, Springer Series in Operations Research and Financial Engineering, https://doi.org/10.1007/978-3-030-12371-0_1

The Markov chains in discrete time are characterized by a finite set $\{1, ..., d\}$, called the *state space*, of possible positions of an object under investigation (particle or agent) and a collection of nonnegative numbers $P = \{P_{ij}\}, i, j \in \{1, ..., d\}$, called the *transition probabilities* from i to j, such that $P_{i1} + \cdots + P_{id} = 1$ for all i. A collection of numbers $P = \{P_{ij}\}$ satisfying these conditions is called a *matrix of transition probabilities* or a *transition matrix* or a *stochastic matrix*. The *Markov chain* specified by the collection P is the process of random transitions between the states $\{1, ..., d\}$ evolving in discrete times $t = 0, 1, 2, \cdots$ by the following rule. If at some time t the object is in a state i, then at the next time step $t + 1$, it migrates from state i to some other state j, choosing j with probability P_{ij}. Then the process repeats starting from j at time $t + 1$, and the same procedure continues ad infinitum.

The Markov chains in continuous time are characterized by a finite set $\{1, ..., d\}$, called the *state space*, of possible positions of an object under investigation (particle or agent), and a collection of nonnegative numbers $Q = \{Q_{ij}\}, i \neq j, i, j \in \{1, ..., d\}$, called the *transition rates* from i to j. To circumvent the restriction $i \neq j$, it is convenient to introduce the negative numbers

$$Q_{ii} = -\sum_{j \neq i} Q_{ij}. \tag{1.1}$$

The positive numbers $|Q_{ii}| = -Q_{ii}$ are called the *intensities of jumps* at state i. The collection $Q = \{Q_{ij}\}, i, j \in \{1, ..., d\}$, is then called a *Q-matrix* or a *Kolmogorov matrix*. The *Markov chain* specified by the collection Q is the process of random transitions between the states $\{1, ..., d\}$ evolving by the following rule. If at some time t the object is in a state i, the object sits at i a random time τ characterized by the following waiting probability:

$$\mathbf{P}(\tau > s) = \exp\{-s|Q_{ii}|\},$$

the random time being referred to as the $|Q_{ii}|$-*exponential waiting time*. At time τ, the object instantaneously migrates from state i to some other state j, choosing j with probability $Q_{ij}/|Q_{ii}|$. Then the process repeats starting from j, and the same procedure continues ad infinitum.

Models with discrete or continuous time are chosen for practical convenience. If we observe a process with some fixed frequency (say, measure a temperature in the sea every morning), the discrete setting is more appropriate. If we observe a process steadily, then the continuous-time setting is more appropriate.

It is useful and popular to represent Markov chains geometrically via oriented graphs. Namely, the *graph of a continuous-time chain with the rates* $Q_{ij}, i, j \in \{1, \cdots, d\}$, is a collection of d vertices with oriented edges e_{ij} (arrows directed from i to j) drawn for the pairs (i, j) such that $Q_{ij} \neq 0$. To complete the picture, the values Q_{ij} can be placed against each edge e_{ij}.

A typical example of a Markov chain with two states is one representing radioactive atoms that can decay after some random time (decay time) or firms that can default after some random time (time to default). In fact, Markov chains are indispensable tools in all branches of science. For instance, to give a few examples, in genetics they model the propagation of various genes through reproduction (which can be randomly chosen from father of mother); in biology they control the processes of evolution; in finance they model the dynamics of prices of financial instruments (say, stocks or options); in economics they are used to model firms' growth and creditworthiness or competition between banks or supermarkets.

The state spaces of Markov chains that will mostly concern us here are the strategies of individuals (say, corrupt or honest inspectors, levels of illegal activity) or the working conditions of agents or devices (say, infected or susceptible computers or individuals, levels of defense).

1.2 Mean-Field Interacting Particle Systems

Let us turn to the basic setting of mean-field interacting particle systems with a finite number of types. We want to model a large number N of particles or agents each evolving according to a Markov chain on some common state space $\{1, ..., d\}$, but with the transition rates depending on the overall distribution of agents among the states. Thus the overall state of the system of such agents is given by a collection of integers $n = (n_1, \cdots, n_d)$, where n_j is the number of agents in the state i, so that $N = n_1 + \cdots + n_d$.

Usually it is more convenient to work with frequencies, that is, instead of n, to use the normalized quantities $x = (x_1, \cdots, x_d) = n/N$, that is, $x_j = n_j/N$. All such vectors x belong to the unit simplex Σ_d, defined as the collection of all d nonnegative numbers summing to unity:

$$\Sigma_d = \{x = (x_1, ..., x_d) : x_j \geq 0 \text{ for all } j \text{ and } \sum_{j=1}^{d} x_j = 1\}.$$

To describe transitions of individuals depending on the overall distribution x, we have to have a family $\{Q(x)\} = \{(Q_{ij})(x)\}$ of Kolmogorov matrices depending on a vector x from Σ_d.

Thus, for every x, the family $\{Q(x)\}$ specifies a Markov chain on the state space $\{1, ..., d\}$. But this is not what we are looking for. We are interested in a system with many agents, i.e., a Markov chain on the states $n = (n_1, \cdots, n_d)$. Namely, the *mean-field interacting particle system* with N agents specified by the family $\{Q(x)\}$ is defined as the Markov chain on the set of collections $n = (n_1, \cdots, n_d)$ with $N = n_1 + \cdots + n_d$ evolving according to the following rule. Starting from any time and current state n, one attaches to each particle a $|Q_{ii}|(n/N)$-exponential random waiting time (where i is the type of this particle). These N times are random, and their minimum is therefore also a random time, which is in fact (as can be shown) a $\sum_i n_i |Q_{ii}(n/N)|$-exponential random waiting time. If this minimal time τ turns out to be attached to some particle of type i, then this particle jumps to a state j

according to the probability law $(Q_{ij}/|Q_{ii}|)(n/N)$. After such a transition, the state n of our system turns to the state n^{ij}, which is the state obtained from n by removing one particle of type i and adding a particle of type j; that is, n_i and n_j are changed to $n_i - 1$ and $n_j + 1$ respectively. After each such transition, the evolution continues starting from the new state n^{ij}, etc.

Equivalently, the transition from a state n can be described by a single clock, as in the standard setting for Markov chains. Namely, to a state n one attaches a $\sum_i n_i |Q_{ii}(n/N)|$-exponential random waiting time. Once the bell rings, one chooses the transition $i \to j, i \neq j$, with probability

$$\mathbf{P}(i \to j) = \frac{n_i Q_{ij}(n/N)}{\sum_k n_k |Q_{kk}(n/N)|}. \tag{1.2}$$

These are well-defined probabilities, since

$$\sum_i \sum_{j \neq i} \mathbf{P}(i \to j) = 1.$$

As was mentioned, as the number of particles N becomes large, the usual analysis of such Markov chains becomes infeasible, a situation often referred to as the "curse of dimensionality." To overcome this problem one searches for a simpler limiting evolution as $N \to \infty$. The situation is quite similar to the basic approaches used in physics to study the dynamics of gases and liquids: instead of following the random behavior of an immense number of individual molecules, one looks for the evolution of the main mean characteristics, such as bulk velocities or temperatures, which often turn out to be described by certain deterministic differential equations.

As will be shown, the limiting evolution of the frequencies $x = (x_1, \cdots, x_d)$ evolving according to the Markov chain described above is governed by the system of ordinary differential equations

$$\dot{x}_k = \sum_{i \neq k} (x_i Q_{ik}(x) - x_k Q_{ki}(x)) = \sum_{i=1}^{d} x_i Q_{ik}(x), \quad k = 1, ..., d \tag{1.3}$$

(the last equation arising from (1.1)), called the *kinetic equations* for the process of interaction described above.

More precisely, under some continuity assumptions on the family $\{Q(x)\}$, one shows that if $X_x^N(t)$ is the position of our mean-field interacting particle system at time t when started at x at the initial time and $X_x(t)$ is the solution of system (1.3) at time t also started at x at the initial time, then the probability for the deviation $|X_x^N(t) - X_x(t)|$ to be larger than any number ϵ tends to zero as $N \to \infty$:

$$\mathbf{P}(|X_x^N(t) - X_x(t)| > \epsilon) \to 0, \quad N \to \infty. \tag{1.4}$$

Let us stress again that system (1.3) is deterministic and finite-dimensional, which is an essential simplification as compared with Markov chains with increasingly large state spaces.

In this most elementary setting the result (1.4) is well known. We will be interested in more general situations in which the transition rates Q are controlled by one or several big players that can influence the dynamics of the chain to their advantage. Moreover, we are concerned with the rates of convergence under mild regularity assumptions on Q. The rates are better characterized in terms of the convergence of some bulk characteristics (so-called weak convergence), rather than directly via the trajectories.

The convergence rates are crucial for any practical calculations. For instance, if the difference between certain characteristics of a Markov chain $X_x^N(t)$ and its limit $X_x(t)$ is of order $1/N$, then applying the approximation $X_x(t)$ to the chain $X_x^N(t)$ yields an error of order 1% for $N = 100$ players and an error of order 10% for $N = 10$ players. Thus the number N does not have to be very large for the approximation (1.3) of "infinitely many players" to give reasonable predictions.

An important question concerns the large-time behavior of mean-field interacting particle systems with finite N. It is usually possible to show that this system spends time t of order the number of particles, $t \sim N$, in a neighborhood of a stable set of fixed points of the dynamics (1.3). Often more precise conclusions can be drawn.

1.3 Migration via Binary Interaction

An important particular case of the scheme above occurs from the setting of *binary interactions*. Namely, let us assume that any player from any state i can be influenced by any player in any other state j to migrate from i to j, the random events occurring with some given intensity $T_{ij}(n/N)/N$ (the appearance of the multiplier $1/N$ being the mark of the general approach according to which each individual's contribution to the overall evolution becomes negligible in the limit of a large number of players N). More precisely, to each ordered pair of agents A, B in different states, say $i \neq j$, one attaches a random $T_{ij}(n/N)/N$-exponential waiting time. The minimum of all these waiting times is again a random waiting time. If the minimum occurs at a pair (A, B) from some states i and j respectively, then A migrates from the state i to the state j, and then the process continues from the state n^{ij}. Equivalently, this process can again be described by one clock with $\sum_{i \neq j} n_i n_j T_{ij}(n/N)/N$-exponential waiting time, attached to a state n. When such a clock rings, the transition $i \to j$, $i \neq j$, is chosen with probability

$$\mathbf{P}(i \to j) = \frac{n_i n_j T_{ij}(n/N)/N}{\sum_{k \neq l} n_k n_l T_{kl}(n/N)/N} = \frac{n_i x_j T_{ij}(x)}{\sum_{k \neq l} n_k x_l T_{kl}(x)}.$$

As is seen directly, the process is the same as the one described above with the matrix $Q_{ij}(x)$ of the type

$$Q_{ij}(x) = x_j T_{ij}(x), \quad i \neq j.$$

Remark 1. *The possibility of such a full reduction of a binary interaction to a mean-field interaction model is due to the simple model of pure migration that we have considered here, and it does not hold for arbitrary binary interactions.*

Remark 2. *Another point to stress is that the ability of any pair to meet is the mathematical expression of full mixing. If that were not assumed (say, spatially separated agents were unable to meet), then the number of interactions would not depend essentially on the size of the population, but rather on the size of a typical neighborhood, and the modeling would be essentially different.*

1.4 Introducing Principals

As was already mentioned, the next step is to include a major player or principal that can influence the evolution of our mean-field interacting particle system. As examples, one can think about the management of complex stochastic systems consisting of large numbers of interacting components: agents, mechanisms, robots, vehicles, subsidiaries, species, police units, etc. To take this into account, our family of $Q(x)$ matrices should become dependent on some parameter b controlled by the principal: $\{Q = Q(x, b)\}$. This can be a real parameter (say, the budget used by the principal for management) or a vector-valued parameter, for instance one specifying resources for various directions of development. In any case, b is supposed to belong to some domain in a Euclidean space, usually bounded, since the resources cannot be infinite.

We shall also assume that the principal has some objectives in this process, aiming to maximize some profit or minimize certain costs arising from the functioning of the Markov chain. In the simplest situation we can assume the principal to be a *best response principal*, which instantaneously chooses the value of b^* maximizing some current profit $B(x, b, N)$ for given x, N:

$$b^* = b^*(x, N) = argmax\, B(x, ., N). \tag{1.5}$$

This procedure reduces the dynamics to the previous case but with $Q^N(x) = Q(x, b^*(x, N))$, which can now depend on N. Convergence (1.4) can still be shown if b^* stabilizes as $N \to \infty$, that is, $b^*(x, N) \to b^*(x)$, as $N \to \infty$, for some function $b^*(x)$. The limiting evolution (1.3) turns into the evolution

$$\dot{x}_k = \sum_{i=1}^{d} x_i\, Q_{ik}(x, b^*(x)), \quad k = 1, ..., d. \tag{1.6}$$

More realistic modeling would require the principal to choose the control parameter strategically with some planning horizon T and a final goal at the end. Moreover, the model should include costs for changing strategies. This leads to a dynamic control problem, whereby the principal chooses controls at fixed moments of time (discrete control setting) or in continuous time (continuous time setting).

Further extensions deal with several principals competing in the background of small players, where the pool of small players can be common for the major agents (think about advertising models, or several states together fighting a terrorist group) or be specific to each major agent (think about generals controlling armies, big banks controlling their subsidiaries, etc.). The development of such models and their dynamic LLN (law of large numbers), leading to more manageable finite-dimensional systems, is the main concern in this book.

1.5 Pressure and Resistance Framework

Let us present now a key particular performance of the above general scheme introduced by one of the authors in [140] and called there the *pressure and resistance game*. In this setting, the interests of the major player P and the small players are essentially opposite, and the mean-field dependence of the actions of small players arises from their binary interactions. Namely, as in the above general setting, the strategies of player P are given by the choices of controls $b(x, N)$, from a given subset of Euclidean space, based on the number of small players N and the overall distribution x of the strategies of the small players. The evolution of the states of small players proceeds as in Section 1.3 with $T_{ij}(x, b)$ depending also on b, so that

$$Q_{ij}(x, b) = x_j T_{ij}(x, b), \quad i \neq j. \tag{1.7}$$

The resulting limiting process described by ODE (1.6) becomes

$$\dot{x}_j = x_j \sum_i x_i [T_{ij}(x, b^*(x)) - T_{ji}(x, b^*(x))], \quad j = 1, ..., d. \tag{1.8}$$

A more concrete performance of this model is the following *pressure and resistance model*. Let us assume that each small player enters on a certain interaction with P (small players try to resist the pressure exerted on them by P) defined by the strategy $j \in \{1, \cdots, d\}$ and that the result of this interaction is the payoff $R_j(x, b)$ to the small player. Moreover, with rate $\varkappa/N$ ($\varkappa$ being some constant), any pair of agents can meet and discuss their payoffs, which may result in the player with lesser payoff R_i switching to the strategy with the better payoff R_j, which may occur with a probability proportional to $(R_j - R_i)$.

In other words, this is exactly the scheme described above by (1.7) with

$$T_{ij}(x, b) = \varkappa (R_j(x, b) - R_i(x, b))^+, \tag{1.9}$$

where we have used the standard notation $y^+ = \max(0, y)$. Therefore, this model is also a performance of the model of Section 1.4 with

$$Q_{ij}(x, b) = \varkappa x_j (R_j(x, b) - R_i(x, b))^+. \tag{1.10}$$

The resulting limiting process described by ODE (1.8) becomes

$$\dot{x}_j = \sum_i \varkappa x_i x_j [R_j(x, b^*(x)) - R_i(x, b^*(x))], \quad j = 1, ..., d. \tag{1.11}$$

Remark 3. *We are working here with a pure myopic behavior for simplicity. The introduction of a random mutation on a global or local level (see, e.g., [124] for standard evolutionary games) would not affect essentially the convergence result below, but would lead to serious changes in the long run of the game, which are worth being exploited.*

Thus we are dealing with a class of models in which a distinguished "big" player exerts a certain level b of pressure on (or interference into the affairs of) a large group of N "small" players that can respond with a strategy j. Often these j strategies are ordered by their strength or level of resistance. The term "small" reflects the idea that the influence of each particular player becomes negligible as $N \to \infty$. As an example of this general setting one can mention the interference of human beings on the environment (say by hunting or fishing) or the use of medications to fight infectious bacteria in the human body, with resisting species having the choice of occupying areas of ample foraging but more dangerous interaction with the big player (large resistance levels r) or less beneficial but also less dangerous areas (low resistance level). Another example can be the level of resistance of a population in a territory occupied by military forces. Some key examples will be discussed in more detail in the next sections.

In many situations, the members of the pool of small players have an alternative class of strategies of collaborating with the big player on various levels. The creation of such possibilities can be considered a strategic action of the major player (who can thus exert some control on the rules of the game). In the biological setting this is, for instance, the strategy of dogs joining humans in hunting their "relatives" wolves or foxes (nicely described poetically as the conversation between a dog and a fox in the famous novel [208]). Historical examples include the strategy of slaves helping their masters to terrorize and torture other slaves and by doing so gaining for themselves more beneficial conditions, as described, e.g., in the classics [33]. As a military example one can cite the strategy of part of the population in a territory occupied by foreign militaries that joins the local support forces for the occupants. For US troops in Iraq, this strategy was discussed in Chapter 2 of [178]. Alternatively, this is also the strategy of a population helping the police to fight criminals and/or terrorists. In the world of organized crime it is also a well-known strategy to play simultaneously both resistance (committing crimes) and collaboration (collaborating with police to eliminate competitors), the classic presentation in fiction being the novel [83].

In general pressure and resistance games, the payoff $R_j(x, b)$ has often the following special features: R increases in j and decreases in b. The dependence of R and b^* on x is more subtle, since it may take into account social norms of various characters. In case of pressure games with resistance and collaboration, the strategic parameter j of small players naturally decomposes into two coordinates $j = (r, c)$, the first one reflecting the level of resistance, and the second the level of collaboration. If the correlation between these activities is not taken into account, the payoff R can be decomposed into the sum of rewards $R = R_j^1(x, b) + R_j^2(x, b)$ with R^1 having the same features as R above, but with R^2 increasing both in j and b.

Another extension to mention is the possibility of having different classes of small players, which brings us to the domain of optimal allocation games, but now in the competitive evolutionary setting, in which the principal (say an inspector) has the task of distributing limited resources as efficiently as possible.

1.6 Nash Equilibria and the Fixed Points of Kinetic Equations

Since the limiting behavior of the controlled Markov chain in the pressure-resistance setting is given by the kinetic equations (1.11), one can expect that eventually the evolution will settle down near some stable equilibrium points of this dynamical system. This remark motivates the analysis of these equilibria.

Moreover, these equilibria have an important game-theoretic interpretation, which is independent of the myopic hypothesis defining the Markov evolution.

Let us consider explicitly the following game Γ_N of $N + 1$ players (which was tacitly borne in mind when we were discussing dynamics). When the major player chooses the strategy b and each of N small players chooses the state i, the major player receives the payoff $B(x, b, N)$ and each player in the state i receives $R_i(x, b)$, $i = 1, \cdots, d$ (as above, with $x = n/N$ and $n = (n_1, \cdots, n_d)$ the realized occupation numbers of all the states). Thus a strategy profile of small players in this game can be specified either by a sequence of N numbers (expressing the choice of the state by each agent), or more succinctly by the resulting collection of frequencies $x = n/N$.

The key game-theoretic notion is that of the *Nash equilibrium*, which is any profile of strategies (x_N, b_N^*) such that for each player, changing its choice unilaterally would not be beneficial. For the game Γ_N, this is a profile of strategies (x_N, b_N^*) such that

$$b_N^* = b_N^*(x_N, N) = argmax\, B(x_N, b, N)$$

and for all $i, j \in \{1, \cdots, d\}$,

$$R_j(x - e_i/N + e_j/N, b_N^*) \le R_i(x, b_N^*). \tag{1.12}$$

Since finding Nash equilibria is a difficult task, one often performs approximate calculations via some numeric schemes leading to approximate Nash equilibria, or ϵ-Nash equilibria. A profile of strategies is an *ϵ-Nash equilibrium* if the above conditions hold up to an additive correction term not exceeding ϵ.

As will be shown, the stationary points of (1.11) describe all approximate Nash equilibria for Γ_N. These approximate Nash equilibria are ϵ-Nash equilibria with ϵ of order $1/N$.

Let us discuss the main concrete examples of the pressure and resistance framework.

1.7 Inspection and Corruption

In the inspection game with a large number of inspectees, see [148], any one from a large group of N inspectees has a number of strategies parametrized by a finite or infinite set of nonnegative numbers r indicating the level at which it chooses to break the regulations ($r = 0$ corresponds to full compliance). These can be levels of tax evasion, levels of illegal traffic through a checkpoint, amounts at which arms production exceeds an agreed level, etc. On the other hand, a specific player, the inspector, tries to identify and punish the trespassers. The inspector's strategies are real numbers b indicating the level of its involvement in the search process, for instance the budget spent on it, which is related in a monotonic way to the probability of the discovery of the illegal behavior of trespassers. The payoff of an inspectee depends on whether its illegal behavior is detected. If social norms are taken into account, this payoff will also depend on the overall crime level of the population, that is, on the probability distribution of inspectees playing different strategies. The payoff of the inspector may depend on the fines collected from detected violators, on the budget spent, and again on the overall crime level (which the inspector may have to report to governmental bodies, say). As time goes by, random pairs of inspectees can communicate in such a way that one inspectee of the pair can start copying the strategy of another one if it turns out to be more beneficial.

A slightly different twist to the model is presented by the entire class of games modeling corruption (see [4, 122, 145, 157, 171] and references therein for a general background). For instance, in developing the initial simple model of [34], a large class of these games studies the strategies of a benevolent principal (representing, say, a governmental body that is interested in the efficient development of the economy) that delegates decision-making power to a malevolent (possibly corrupt) agent, whose behavior (legal or not) depends on the incentives designed by the principal. The agent can deal, for example, with tax collection from private enterprises, which can use bribes to persuade a corrupt tax collector to accept falsified revenue reports. In this model, the set of inspectors can be considered to be a large group of small players that can choose the level of corruption (quite in contrast to the classical model of inspection) by taking no bribes at all, or not too much in the way of bribes, etc. The strategy of the principal consists in fiddling with two instruments: choosing wages for inspectors (to be attractive enough that the agents will be afraid to forfeit them) and investing in activities aimed at the timely detection of fraudulent behavior. Mathematically, these two types are fully analogous to the preemptive and defensive methods discussed in models on counterterrorism; see below.

In the standard setting of inspection games with a possibly tax-evading inspectee (analyzed in detail in [148] under some particular assumptions), the payoff R looks as follows:

$$R_j(x, b) = r + (1 - p_j(x, b))r_j - p_j(x, b)f(r_j), \tag{1.13}$$

where r is the legal payoff of an inspectee, various r_j denote various amounts of undeclared profit, $j = 1, \cdots, d$, $p_j(x, b)$ is the probability of the illegal behavior of an inspectee being discovered when the inspector uses budget b for the searching operation, and $f(r_j)$ is the fine that the guilty inspectee has to pay on being discovered.

In the standard model of corruption "with benevolent principal," see, e.g., [4], one sets the payoff of a possibly corrupt inspector (now taking the role of a small player) as

$$(1 - p)(r + w) + p(w_0 - f),$$

where r is now the bribe an inspector asks from a firm to agree not to publicize its profit (and thus allowing the inspectee not to pay its taxes), w is the wage of an inspector, f the fine the inspector has to pay when its corruption is discovered, and p is the probability of corrupt behavior being discovered by the benevolent principal (say, governmental official). Finally, it is assumed that when the corrupt behavior is discovered, the agent not only pays a fine, but is also fired from its job and has to accept a lower-level employment with reservation wage w_0. In our strategic model we denote by r the strategy of an inspector with possible levels $r_1, \cdots, r_d$ (the amount of bribes it is taking), and the probability p of discovery will depend on the effort (say, budget b) of the principal and the overall level of corruption x, with a fine depending on the level of illegal behavior. This natural extension of the standard model leads to the payoff

$$R_j(x, b) = (1 - p_j(x, b))(r_j + w) + p_j(x, b)(w_0 - f(r_j)), \tag{1.14}$$

which is essentially identical to (1.13).

1.8 Cybersecurity, Biological Attack–Defense, and Counterterrorism Modeling

A "linguistic twist" that changes "detected agents" into "infected agents" brings us directly to the (seemingly quite different) setting of cybersecurity or biological attack–defense games.

For instance, let us look at applications to botnet defense (for example, against the famous Conficker botnet), widely discussed in the contemporary literature, since botnets (zombie networks) are considered to pose the biggest threat to international cybersecurity, see, e.g., the review of the abundant bibliography in [41]. The comprehensive game-theoretic framework of [41] (which extends several previous simplified models) models a group of users subject to a cybercriminal attack of botnet herders as a differential game of two players, the group of cybercriminals and the group of defenders.

Our approach adds networking aspects to this analysis by allowing the defenders to communicate and eventually copy more beneficial strategies. More concretely, our general model of inspection or corruption becomes almost directly applicable in this setting by the clever linguistic change of "detected" to "infected" and by considering the cybercriminal the "principal agent"! Namely, let r_j (the index j being taken from some discrete set here, though the more advanced theory of the next sections allows for a continuous parameter j) denote the level of defense applied by an individual (computer owner) against botnet herders (an analogue of the parameter γ of [41]), which can be the level of antivirus programs installed or the measures envisaged to quickly report and repair a problem once detected (or possibly a multidimensional parameter reflecting several defense measures). Similarly to our previous models, let $p_j(x, b)$ denote the probability of a computer being infected given the level of defense measures r_j, the effort level b of the herder (say, budget or time spent), and the overall distribution x of infected machines (this "mean-field" parameter is crucial in the present setting, since infection propagates as a kind of epidemic). Then, for a player with a strategy j, the cost of being (inevitably) involved in the conflict can be naturally estimated by the formula

$$R_j(x, b) = p_j(x, b)c + r_j, \qquad (1.15)$$

where c is the cost (inevitable losses) of being infected (thus one should aim at minimizing this R_j, rather than maximizing it as in our previous models). Of course, one can extend the model to various classes of customers (or various classes of computers) for which values of c or r_j may vary and by taking into account more concrete mechanisms of virus spreading, as described, e.g., in [164, 167].

Similar models can be applied to the analysis of defense against a biological weapon, for instance by adding the active agent (principal interested in spreading the disease) into the general mean-field epidemic model of [165], which extends the well-established SIS (susceptible–infectious–susceptible) and SIR (susceptible–infectious–recovered) models.

Yet another set of examples represents models of terrorist attacks and counterterrorism measures; see, e.g., [13, 52, 211, 212] for the general background on game-theoretic models of terrorism, and [81] for more recent developments. We again suggest here a natural extension to basic models that provides the possibility of a large number of interacting players and various levels of attacks, the latter extension being in line with arguments from [65] advocating consideration of "spectacular attacks" as part of a continuous scale of attacks of various levels.

In the literature, counterterrorist measures are usually decomposed into two groups, so-called proactive (or preemptive), such as direct retaliation against a state sponsor, and defensive (also referred to as deterrence), such as strengthening security at an airport, with the choice between the two considered the main strategic parameter. As stressed in [202], the first group of actions is "characterized in the literature as a pure public good, because a weakened terrorist group poses less of a threat to all potential targets," but on the other hand, it "may have a downside by creating more grievances in reaction to heavy-handed tactics or unintended collateral damage" (because it means to "bomb alleged terrorist assets, hold suspects without charging them, assassinate suspected terrorists, curb civil freedoms, or impose retribution on alleged sponsors"), which may result in the increase of recruitment of terrorists. Thus, the model of [202] includes the recruitment benefits of terrorists as a positively correlated function of preemption efforts.

A direct extension of the model of [202] along the lines indicated above (large number of players and levels of attacks) suggests that one write down the reward of a terrorist, or a terrorist group, considered as a representative of a large number of small players, using one of the levels of attack $j = 1, \cdots, d$ (in [202] there are two levels, normal and spectacular only), as

$$R_j(x, b) = (1 - p_j(x, b))r_j^{fail}(b) + p_j(x, b)(S_j + r_j^{succ}(b)), \qquad (1.16)$$

where $p_j(x, b)$ is the probability of a successful attack (which depends on the level b of preemptive efforts of the principal and the total distribution of terrorists playing different strategies), S_j is the direct benefits in case of success, and $r_j^{fail}(b), r_j^{succ}(b)$ are the recruitment benefits in the cases of failure or success respectively. The costs of the principal are given by

$$B(x, b) = \sum_j x_j \left[(1 - p_j(x, b))b + p_j(b)(b + S_j) \right].$$

It is seen directly that we are again in the same situation as described by (1.14) (up to constants and notation). The model extends naturally to account for the possibility of actions of two types, preemption and deterrence. Also very natural would be the extension to several major players For instance, USA and EU cooperation in fighting terrorism was considered in [13].

1.9 Coalition Building, Merging and Splitting, Network Growth, Queues

In a large number of problems it is essential to work with an infinite state space of small players, in particular, with the state space of all natural numbers. Mathematical results are much rarer for this case than with finite state spaces. This infinite-dimensional setting is crucial for the analysis of models with growth, such as merging banks or firms (see [196, 206]) or the evolution of species and the development of networks with preferential attachment (the term coined in [27]), for instance scientific citation networks or the network of internet links (see a detailed discussion in [155]).

Apart from the obvious economic examples mentioned above, a similar process in the growth of coalitions under pressure can possibly be used for modeling the development of human cooperation (forming coalitions under the "pressure" exerted by nature) or the creation of liberation armies (from initially small guerrilla groups) by the population of the territories oppressed by an external military force.

Yet another example can be represented by the analysis of the financial support of various projects (developmental or scientific) when the funding body has tools to merge or split projects and transfer funds between them over certain periods of time.

Our evolutionary theory of coalition-building (properly developed in Chapter 4) can be considered an alternative to the well-known theory of coalitional bargaining; see, e.g., [64].

Models of growth are known to lead to power laws in equilibrium, which are verified in a variety of real-life processes; see, e.g., [206] for a general overview and [199] for particular applications to crime rates. In Chapter 4 we will deal with the response of a complex system that includes both the capacity for growth and coalition-building mechanisms to external parameters that may be set by the principal (say, by governmental regulations), that has its own agenda (may wish to influence the growth of certain economics sectors). The modern literature on network growth is very extensive, and we will not try to review it, referring for this purpose, for instance, to [44] and references therein.

Of course, these processes of growth have clear physical analogues, such as the formation of dimers and trimers by molecules of gas with eventual condensation under (now real physical) pressure. The relation with Bose–Einstein condensation is also well known; see, e.g., [216].

We shall pay attention only to infinite spaces that are countable, which are possibly mostly relevant for practical studies of evolutionary growth. For this class of models the number N of agents becomes a variable (and is usually increasing as a result of evolution), and the major characteristic of the system becomes just the distribution $x = (x_1, x_2, \cdots)$ of the sizes of the groups. The analysis of the evolution of these models has a long history, see [216], though in the abundant literature on models of evolutionary growth, the discussion usually starts directly with the deterministic limiting model, with the underlying Markov model just mentioned as a motivating heuristic. Mathematically, the analysis is similar to that of finite state spaces, though serious technical complications may arise. We develop the "strategically enhanced model" in Section 4.5, analyzing such evolutions under the "pressure" of strategically varying parameters set by the principal.

As another insightful example of games with a countable state space let us mention the model of the propagation of knowledge from [60, 77]. In this model an agent is characterized by a number $n \in \mathbf{N}$ that designates the amount of information the agent has about a certain subject. Meetings of two agents with information n and m occur with certain intensities (which represent the control parameters) and result in each of them acquiring the total information $m + n$. The summation of the information of agents is based, of course, on a highly idealized assumption of independent knowledge. In any case, in contrast to the model of merging, the number of agents here remains constant, but their information (the analogue of mass) is increasing.

Yet another large class of models with countable state spaces comprises various models of queues parametrized by the sizes of the queues at different servers; see, e.g., [91, 92] and references therein for detail.

1.10 The Minority Game, Its Strategic Enhancement, and the Sex Ratio Game

An important example of the pressure and resistance framework is represented by the famous minority game designed to model the evolution of financial markets. The remarkable properties of the minority game were revealed partly by extensive computer simulations and partly by an application of the methods of statistical mechanics.

Recall that in the *minority game*, each of N players declares (or adopts) independently and simultaneously one of the two possible strategies, denoted by 0 and 1. The payoff of 1 unit is paid to all players that declared the strategy that turned out to be in the minority, and the other players pay a fine of 1 unit. In the rare event of equal numbers having declared each of the two strategies, all payoffs can be set to zero. The game is supposed to be repeated many times, so that players can learn to apply better strategies.

Let n_0, n_1 denote the numbers of players that have chosen 0 or 1 respectively in one step of the game. Then $n_0 + n_1 = N$. Let $x = n_0/N$. We are exactly in the situation described in Sections 1.5, 1.6 (without the major player) with $d = 2$ and payoffs

$$R_1(x) = -R_0(x) = \mathrm{sgn}\,(x - 1/2) = \begin{cases} 1, & x > 1/2, \\ 0, & x = 1/2, \\ -1, & x < 1/2. \end{cases} \tag{1.17}$$

The only specific feature of this setting is the obvious discontinuity of the payoffs $R_j(x)$. However, for large N, the chance of a draw is negligible, so that the model will not be changed essentially if one takes a smooth version of the payoffs,

$$R_1(x) = -R_0(x) = \theta(x - 1/2), \tag{1.18}$$

where $\theta(x)$ is a smooth nondecreasing function such that $\theta(x) = \text{sgn}(x)$ for $|x| \geq \epsilon$ with some small ϵ. The kinetic equations (1.3) can be written down in terms of a single variable x as

$$\dot{x} = -x(1-x)\theta(x - 1/2). \tag{1.19}$$

An analysis of the dynamics arising from this game is given in Sections 2.12 and 2.13.

Placing the minority game in this general setting suggests various reasonable extensions of this market model, for instance by including major players (say, market makers) that have tools to control the interactions or the payoffs of minor players, or by a more subtle grouping of players. Moreover, it reveals a nice link with the so-called *sex-ratio game* of evolutionary biology. The latter game is designed to explain the common $1/2$ probability for births of males and females in many populations. The rules of the sex-ratio game arise from the observation that if there are n_0 males and n_1 females in a population (its current generation), then producing a male (respectively female) offspring would yield to this offspring an expected fraction n_1/n_0 (respectively n_0/n_1) of its own offspring. Thus we find ourselves in the situation described by the rules above with payoff (1.18) having the form

$$R_1(x) = \theta(x - 1/2) = x/(1-x). \tag{1.20}$$

1.11 Threshold Collective Behavior

A similar two-state model arises in a seemingly different setting of the models of *threshold collective behavior*, for example, the well-known model of mob control (see [53] for a general introduction). It is assumed there that a crowd consists of a large number N of agents that can be in two states, excited (say, aggressive) and quiet, denoted by 1 and 0 respectively. In discrete-time modeling, the excitation is supposed to propagate by an agent copying its neighbors. Namely, each agent i is characterized by the threshold level $\theta_i \in [0, 1]$ and the influence vector $w_{ji} \geq 0$, $j \neq i$, where w_{ji} is the weight of the opinion of i for agent j. If at some time t the states of the agents are given by the profile vector $\{y_1^t, \cdots, y_N^t\}$, $y_j^t \in \{0, 1\}$, then an agent i in the next moment of time $t + 1$ becomes excited if the weighted level of excitation of its neighbors exceeds the threshold θ_i, that is, y_i^{t+1} turns into 1 or 0 depending on whether or not $\sum_{j \neq i} w_{ij} y_j \geq \theta_i$. Thus each agent acts as a classical *perceptron* (threshold decision unit) of the theory of neural networks. Quite similar are more general models of the so-called *active network structures* that are used to model the propagation of opinion in social media such as Facebook and Twitter.

Concerning the θ_j, it is often assumed that they are independent identically distributed random variables with some distribution function $F(x) = \mathbf{P}(\theta_i \leq x)$. To fit this process to our general model it is necessary to assume that the mob is decomposed into a finite number of classes with identical weights, so that w_{ij} can be characterized as the weights of influence between representatives of different classes, rather than individual agents, and in fact, the theory usually develops under an assumption of this kind. The standard simplifying assumption is the so-called *anonymity case*, whereby all influences are considered to be identical, that is, $w_{ij} = 1/(N - 1)$ for all $i \neq j$. Let us assume that such is the case. One can model the propagation of excitation in either discrete or continuous time. For definiteness, let us choose the second approach (more in line with our general modeling). Then the propagation is measured in terms of the rates of transitions, rather than probabilities, and we naturally assume here that the rate of transition from state 0 to state 1 is proportional to $\mathbf{P}(\theta_i \leq x) = F(x)$, and that the rate from state 1 to state 0 is proportional to $\mathbf{P}(\theta_i > x) = 1 - F(x)$, where $x \in [0, 1]$ is the proportion of the excited states among the agents. Thus we find ourselves fully in our general setting with a two-state model, $d = 2$, and the Kolmogorov matrix $Q_{ij}, i, j \in \{0, 1\}$, with

$$Q_{01} = \alpha F(x), \quad Q_{10} = \alpha(1 - F(x)),$$

with some constant α, and thus $Q_{00} = -\alpha F(x)$, $Q_{11} = -\alpha(1 - F(x))$. A constant α can be used to extend the model to the case that every agent is influenced only by its neighbors, where α can be characterized by the average number of these neighbors.

As for any model with $d = 2$, the kinetic equations (1.3) can be written in terms of just one equation for x:

$$\dot{x} = xQ_{11} + (1 - x)Q_{01} = -x\alpha(1 - F(x)) + (1 - x)\alpha F(x) = \alpha(F(x) - x),$$

that is, finally, as

$$\dot{x} = \alpha(F(x) - x), \tag{1.21}$$

which is a standard equation used in modeling threshold collective behavior (see [53]).

1.12 Project Selection, Investment Policies, and Optimal Allocation

As yet another example let us introduce a simple model for *project selection*. The specific feature of this model is the necessity to work with a two-parameter family of states. Assume an that investment body aims at developing several directions $j \in \{1, \cdots, d\}$ of business or science, ordered according to their importance, with some budget allocated to each of these directions. When a call for bidding for the support of projects is declared, the interested parties have to choose both the project to bid for and the level of investment in their efforts in preparation of the bid, which will have some possible costs $c \in \{c_1, \cdots, c_m\}$, with $c_1 < \cdots < c_m$. The probability of a project being selected, $p(j, c, x_j)$, will be thus an increasing function of j, an increasing function of c, and a decreasing function of $x_j = n_j/N$, the fraction of bids in the jth direction. Therefore, if A_j is an award (the value of a grant) for a project in the jth direction, then the average payoff of small players can be written as

$$R_{j,c}(x) = -c + A_j p(j, c, x_j). \tag{1.22}$$

The kinetic equations, see (1.3) and (1.10), in this two-dimensional index model take the form

$$\dot{x}_{jc} = \sum_{i,\tilde{c}} \varkappa x_{i\tilde{c}} x_{jc} (R_{i,\tilde{c}}(x) - R_{j,c}(x)). \tag{1.23}$$

Here both A_j and the functions $p(j, c, x)$ are the parts of the mechanism design of the principal (investment body). They can be encoded, of course, by certain parameters b that are chosen by the principal.

The model is a performance of the problem of optimal allocation in a competitive environment. It can be extended in various directions, for instance, by taking into account the initial expertise of the bidders and the related costs of switching between the topics.

1.13   Games on Networks and Positivity-Preserving Evolutions

Here we stress two rather trivial aspects of evolution (1.3). The first is its universality as a positivity-preserving evolution. Namely, it is not difficult to show (see, e.g., [136], Sections 1.1 and 6.8) that every sufficiently regular transformation of $\mathbf{R}^d$ preserving positivity and the "number of particles" $x_1 + \cdots + x_n$, in other words every smooth transformation of the simplex Σ_d, can be described by an ODE of type (1.3). This is, in fact, a performance of a more general representation result for positivity-preserving transformations of measures.

The second aspect is just a geometric rewording. Following the general idea of the graphical representation of Markov chains (see Section 1.1), we can similarly draw a graph of the mean-field interacting particle system of Section 1.2, which will be a graph with d vertices and edges e_{ij} drawn whenever $Q_{ij}(x)$ does not vanish identically. The evolution (1.3) can then be looked at as the evolution of the "weights" of the vertices. Due to this geometric interpretation, mean-field interacting particle systems and their limits (1.3) are sometimes referred to as *networked models*; see, e.g., [188] for a concrete epidemic model presented in this way.

1.14   Swarm Intelligence and Swarm-Based Robotics

Swarm-based robotics is often understood as an attempt to build algorithms or mechanisms for the functioning of complex systems of interacting robots by analogy with analagous observed behavior in nature, mostly social insects such as ants and bees.

The evolutions of such mechanisms are usually represented as mean-field interacting particle systems, thus given approximately (in the LLN limit) by kinetic equations (positivity-preserving evolutions on the simplexes; see Section 1.13 above). A typical example (taken from [48]) is as follows. Two types of workers (for instance, so-called minor and major in some ant species) of quantity n_1 and n_2 respectively, with total number $N = n_1 + n_2$, can perform a certain task (say, clean the nest of dead bodies). The level of unfulfilled task s increases in time during the functioning of the nest and decreases proportionally to the number of workers engaging in it:

$$\dot{s} = \delta - \alpha(N_1 + N_2)/N, \tag{1.24}$$

where N_i is the number of workers of type i performing the job, and δ, α are some positive constants. The workers of the two types have propensities or inclinations to do the task $T_1(s)$ and $T_2(s)$ (mathematically, the rates with which they take it on), depending on the level of stimulus specified by s (say, difficulty to move in a nest full of rubbish). If p is the probability of giving up the job after starting it, the evolution of the frequencies $x_i = N_i/n_i$ is given by the (easy to interpret) equations

$$\dot{x}_1 = T_1(s)(1 - x_1) - px_1, \quad \dot{x}_2 = T_2(s)(1 - x_2) - px_2. \tag{1.25}$$

For system (1.24)–(1.25), the stationary points can be found explicitly. More general situations (with several tasks and several types of workers) have been analyzed numerically, giving reasonable agreement with the concrete observations of ants, and then were used to build algorithms for robots.

A well-known analytic model of collective behavior of a swarm of interacting Brownian oscillators was proposed by Y. Kuramoto; see [1] for a review and extensions. As was shown in [233], the corresponding dynamics can be also obtained as the solution to a certain mean-field game.

The theory developed in this book offers the precise relations between the processes with a finite number of agents and their LLN limits, like (1.24)–(1.25). Moreover, it suggests choosing the parameters available to the major player (say, the designer of the swarm) strategically via forward-looking perspectives with finite or infinite horizon (in the spirit of Chapter 3).

1.15 Replicator Dynamics with Mean-Field Dependent Payoffs: Controlled Hawk-and-Dove Evolution

As a final example let us touch upon applications to the well-established theory of evolutionary games and evolutionary dynamics, reducing the discussion to the variations on the most popular concrete model.

Recall the celebrated *hawk-and-dove game*, which is a symmetric two-player game with the matrix of payoffs (the letters in the table denote the payoffs to the row player)

	hawk	dove
hawk	H	V
dove	0	D

The game can be interpreted as being played by individuals sharing a common resource. If a hawk and a dove meet, the dove abandons the site without a fight, leaving the whole resource, V, to the hawk. If two doves meet, they share the resource, yielding D each (which could be more or less than V, depending on the level of collaboration). If two hawks meet, they fight, inflicting considerable damage on each other, so that the average payoff can become even negative.

The *replicator dynamics*, which is one of the most basic dynamics for the theory of evolutionary games (see [144] or [115] for general background) is given by the equation

$$\dot{x} = x(1 - x)[x(H + D - V) - (D - V)],$$

which is of course a particular case of dynamics (1.3), (1.10). This evolution may have only one internal stationary point (point of internal equilibria),

$$x^* = \frac{V - D}{V - D - H}.$$

If $V > D$, as is usually assumed, then $x^* \in (0, 1)$ if and only if $H < 0$. In this case, this point is known (and easily checked) to be a stable equilibrium of the replicator dynamics.

Remark 4. *This equilibrium is also known to be an ESS (evolutionary stable equilibrium) of the hawk-and-dove game (see, e.g., [144] or [115] for the general background on ESS).*

The replicator dynamics can be obtained as the dynamic LLN (law of large numbers) for a Markov model of N interacting species, which can use one of two strategies, h or d (hawk or dove), with each pair meeting randomly and playing the game of given by the table above. The payoff for each player in this game of N players, the fraction x (respectively $1 - x$) for those playing h (respectively d), is $Hx + V(1 - x)$ for being a hawk and $D(1 - x)$ for being a dove.

The point we like to stress here is the following. If there are only a few hawks and plenty of dove to rob, the hawks have less impetus to fight among themselves, meaning that H may increase with the decay of x, the fraction of the species playing h. Similarly, if there are too many doves in the restricted resource area, they may begin behaving more aggressively, meaning that D may decrease with the decay of x. Choosing linear functions to express this dependence in the simplest form, the modified table of this *nonlinear hawk-and-dove game* becomes

	hawk	dove
hawk	$H + ax$	V
dove	0	$D - bx$

with some constants $a, b > 0$.

In this setting, the game of two players is no longer defined independently, but the game of N players choosing one of the strategies h and d remains perfectly well defined (with the theory mentioned in Section 1.6 and developed in Chapters 2, 3 fully applied), with the payoffs of hawks and doves becoming

$$\Pi_h(x) = (H + ax)x + V(1 - x), \quad \Pi_d(x) = (D - bx)(1 - x).$$

The replicator dynamics generalizes to the equation

$$\dot{x} = x(1 - x)[x(H + ax + D - bx - V) - (D - bx - V)], \tag{1.26}$$

which remains a particular case of dynamics (1.3), (1.10). The internal fixed points of this dynamics are the solutions of the quadratic equation

$$(a - b)x^2 - (V - D - H - b)x + (V - D) = 0. \tag{1.27}$$

It is seen that, say for $a = b$, the internal equilibrium becomes

$$x^* = \frac{V - D}{V - D - H - b},$$

which moves to the left with increasing b. When the denominator in this expression becomes negative, the internal equilibrium gets lost, and the pure strategy d becomes the stable equilibrium. On the other hand, if $b = 0$, then the roots of the quadratic equation (1.27) have asymptotic expressions

$$x_1 = \frac{V - D}{V - D - H}\left(1 + a\frac{V - D}{(V - D - H)^2} + O(a^2)\right), \quad x_2 = \frac{1}{a}(V - D - H) + O(1)$$

for small a. Therefore, for small a, the internal stationary point begins moving to the right, remaining the only internal stationary point. When a increases, various bifurcations can occur. For instance, if $-H < a < (V - D - H)/2$, the internal stationary points disappear. If

$$\frac{V - D - H}{2} < a < \frac{(V - D - H)^2}{4(V - D)},$$

then two internal fixed points appear (both solutions of (1.27) belong to $(0, 1)$), the left being stable, and the right unstable, so that in total, we get two stable equilibria, one mixed and one pure hawk.

Similar shifts of equilibria and bifurcations occur when mean-field dependence is introduced in other popular two-player games. It would be interesting to see whether this kind of bifurcations can be observed in real field experiments.

The model can be used to describe the transformations of behavioral patterns and social norms in society, how doves (social behavior) develop themselves into hawks (aggressive behavior, violence) and vice versa. Moreover, making the parameters a, b depending on the some control parameter u of a major player allows one to introduce the model of controlled hawk-and-dove evolution, with the principal having a tool to drive the equilibria in the direction desired. Results of Chapter 3 provide a general technique for the analysis of such controlled processes.

1.16   Fluctuations

An important direction that we are not pursuing here is the analysis of fluctuations of the random Markov processes of N agents and the deterministic LLN provided by the kinetic equations. This problem can be approached in two slightly different ways. First, one can begin with the generator of the N agent process and look for the next order of its approximation (as compared with the first-order approximation used for obtaining the kinetic equations). Expanding the increments of functions to the second order yields directly the second-order operator, and hence the diffusion process. Such diffusion processes may capture much better the long-time behavior of the N-agent process, for instance implying the dying out of certain strategies, which has been demonstrated both theoretically and experimentally on many concrete biological systems with the evolution described by the models of evolutionary games; see [74] and references therein. For models with countable state space, this analysis was performed in [129]. An alternative approach is to write down explicitly the parameters (e.g., the generator) of the process of fluctuations and look for its limit under appropriate scaling. Such limits can be often found to be described by a certain Gaussian process, thus leading to *dynamic central-limit-type* theorems. Such results have been obtained for various models; see the references in Appendix. Both approaches can be looked at as providing a stochastic law of large numbers for processes involving N agents. Unlike deterministic LLN, these random LLN are less robust and much more sensitive to the methods of approximations and scaling, with a variety of limits that can be obtained depending on the choice of the latter.

1.17   Brief Guide to the Content of This Book

As was already mentioned, the first part of this book deals with models of a large number N of minor players interacting with major players, the strategies of small players propagating by direct imitation (myopic behavior) of more successful strategies of the surrounding players. The purpose of the analysis is to turn the "curse of dimensionality" into the "blessing of dimensionality" by obtaining more manageable deterministic limits, that is, kinetic equations, as $N \to \infty$, and providing effective rates of convergence. Numerous examples of concrete models were presented above.

The first part is mostly based on papers [138, 140], with a gap from [138] corrected (see Remark 17) and with the results of [140] improved and strengthened in many ways, specifically for countable state spaces.

In Chapter 2 we shall work with the case of a finite number d of strategies of small players and major players acting with the instantaneous best response. Together with convergence rates to the deterministic limit, we present the game-theoretic interpretation of the fixed points of the limiting dynamics, providing a link between the large-time behavior of the limiting, $N \to \infty$, and approximating, finite N, dynamics. The results of the last Sections 2.11–2.14 are essentially new, though the last section can be looked at as a performance of the general results from [133] or [177] in the specific setting of interacting particles.

In Chapter 3 the behavior of major players is modified by the more realistic requirement that they choose their control parameters strategically, aiming at some optimal payoff during either a fixed finite period or over an infinite horizon. The dynamics is considered in the multistep version, in which major players choose their controls at discrete times $k\tau$, $k \in \mathbf{N}$, with a fixed τ, and in the continuous-time limit, obtained from the former by letting $\tau \to 0$. The deterministic continuous-time problem is obtained when $\tau \to 0$ and $N \to \infty$ are simultaneously subject to some bounds on the dependence of τ and N. Long-time behavior is analyzed both for the discounted payoff and for the average payoff, the latter often leading to the so-called turnpike behavior of the trajectories, well known in classical mathematical economics.

Chapter 4 extends the results of Chapters 2 and 3 to the case of countable state spaces of small players. This extension is carried out in order to include important models of evolutionary coalition-building and strategically enhanced preferential attachment. The mathematics of this chapter is more demanding, since it heavily exploits the theory of infinite-dimensional ODEs. The results of Section 4.6 are new.

In the second part we model small players as rational optimizers, bringing the analysis into the realm of so-called mean-field games (MFG), widely discussed in the modern research literature. We begin in Chapter 6 with a general introduction to MFGs on finite state spaces, including the theorem on the approximations of finite-number-of-player games by limiting MFG evolutions, following mostly the papers [29, 30], but with significant improvements and simplifications.

The rest of the second part is based essentially on papers [126, 143, 145, 146]. Since the complexity of MFGs increases with an increase in the size of the state space of small players, we begin with detailed description of toy models of three-state and four-state games in Chapters 6 and 7 respectively. These games represent remarkable and rare examples, in which all stationary solutions to the MFG consistency problem can be found explicitly. The analysis of these toy models reveals and provides an explicit description of the nontrivial properties of MFGs, such as phase transitions, and of the key parameters that control these transitions. It also represents a reasonably large class of models, in which the state space of small players is two-dimensional, one direction being controlled by the individual player (say, the level of tax evasion or the costs spent on a defense system) and another by the principal or binary interactions (say, the position of an agent on the bureaucratic ladder or the state of a computer being infected or not).

In Chapter 8 a more general theory is developed, revealing the turnpike property in the MFG context. Roughly speaking, the turnpike property means that time-dependent control evolutions with large horizon spend most of their time around some fixed stationary solution (a turnpike), which leads to the crucial simplification of the analysis, because stationary solutions are usually much easier to identify and calculate.

A very detailed analytic description of models is not achievable for models with larger state spaces. For such models, numeric calculations can be effectively exploited for obtaining a detailed picture of the dynamics, its phase portrait, and bifurcations. We are not touching on numerics in this book. Instead of numerics, we employ asymptotic analysis to describe the dynamics under reasonable asymptotic regimes. This approach is akin to that in physics whereby the search for a relevant small parameter is the key tool for obtaining insights into a complicated phenomenon. We identify and exploit three natural small parameters in our MFGs, leading to the following three asymptotic regimes: fast execution of personal decisions (why should one wait long to execute one's own decisions?), small discounts (this means in practical terms that the planning horizon is not very small), and small coupling constants of binary interaction, the last one being, of course, the most standard one in physics.

Finally, we present a literature review, indicating the main trends both in our main methodology (dynamic law of large numbers) and in the concrete areas of applications (inspection, corruption, etc.) most closely related to our methods and objectives. The literature is quite abundant and is growing rapidly.

Part I
Multi-agent Interaction and Nonlinear Markov Games

Chapter 2
Best-Response Principals

In this chapter our basic tool, the dynamic law of large numbers (LLN, whose descriptive explanation was given in Chapter 1), is set on a firm mathematical foundation. We prove several versions of this LLN, with different regularity assumptions on the coefficients, with and without major players, and finally with a distinguished (or tagged) player, the latter version being used later, in Part II. convergence is proved with rather precise estimates of the error terms, which is, of course, crucial for any practical applications. For instance, we show that in the case of smooth coefficients, the convergence rates, measuring the difference between the various bulk characteristics of the dynamics of N players and the limiting evolution (corresponding to an infinite number of players), are of order $1/N$. For example, for $N = 10$ players, this difference is about 10%, showing that the number N does not have to be "very large" for the approximation (1.3) of "infinitely many players" to give reasonable predictions.

In the presence of a major player, the theory is built here for the model of what can be called the *best-response principle*, whereby the major player chooses the strategy at any time as the instantaneous best response (in case of several major players as an instantaneous Nash equilibrium) for the current distribution of small players. More realistic forward-looking major players will be analyzed in the next chapter.

Further in this chapter we develop the theory of approximate Nash equilibria for our mean-field interacting systems of a large number of agents, showing that these equilibria can be approximated by the stationary points of the system of the limiting kinetic equations. This result yields a remarkable static game-theoretic interpretation of the stationary points of kinetic equations, which is independent of the myopic hypothesis and migration models used in the derivation of the latter. When these stationary points are dynamically stable, we show further that the Markov evolution of the systems of many agents spends large periods of time in a neighborhood of these stationary points. Moreover, in some cases, the connection becomes even deeper: the supports of the invariant measures of the Markov evolutions of N agents turn out to

© Springer Nature Switzerland AG 2019

V. N. Kolokoltsov and O. A. Malafeyev, *Many Agent Games in Socio-economic Systems: Corruption, Inspection, Coalition Building, Network Growth, Security*, Springer Series in Operations Research and Financial Engineering, https://doi.org/10.1007/978-3-030-12371-0_2

be concentrated around these fixed points. This result makes the crucial extension of the LLN limit from finite to infinite times.

Next we extend our modeling to the case of discontinuous coefficients linking uniqueness and nonuniqueness of the solutions to the kinetic equations with the problem of the existence of the invariant measures for the prelimiting game of a finite number of agents. Discontinuous coefficients arise naturally when small payers are considered rational optimizers (that is, in the setting of MFG) with a finite-valued control parameter, with switching between these values creating discontinuities, separating regular regimes.

Finally, we extend the modeling to the so-called semi-Markov chains, allowing for the waiting times between the jumps to be nonexponential but to have heavy power tails. In this case, the limiting dynamics turns into a fractional PDE, with the solution process obtained by a certain random-time change of the solutions of the initial kinetic equations. In particular, the dynamic LLN limit becomes a random process, unlike all other situations discussed in this book. This brings the theory into the realm of the fractional calculus, which becomes one of the major trends in the modern application of analysis to physics and economics. In fact, many observations of naturally occurring processes give strong evidence of the presence of heavy tails in the distribution of time between various transitions (see, e.g., [222] or [229]).

2.1   Preliminaries: Markov Chains

Two basic notions that are crucial for our analysis are Markov chains and the characteristics of first-order linear partial differential equations (PDE). We briefly recall now these notions (with more analytic detail than in Section 1.1), referring for a systematic exposition to numerous textbooks; see, e.g., [185] or [113] for Markov chains and [142, 192, 193] for ODEs and PDEs.

Let us first recall the basic setting of continuous-time Markov chains with a finite state space $\{1, ..., d\}$, which can be interpreted as types of agents or particles (say, possible opinions of individuals on a certain subject, or the levels of fitness in a military unit, or the types of robots in a robot swarm). A Markov chain on $\{1, ..., d\}$ is specified by the choice of a *Q-matrix* or a *Kolmogorov matrix Q*, which is a $d \times d$ square matrix such that its nondiagonal elements are nonnegative and the elements of each row sum to zero (and thus the diagonal elements are nonpositive). As was mentioned in Chapter 1, a *Markov chain* with Q-matrix Q is a process evolving by the following rule. Starting from any time t and a current state i, one waits a $|Q_{ii}|$-exponential random waiting time τ, and then the position jumps to a state j according to the distribution $Q_{ij}/|Q_{ii}|$. Then at time $t + \tau$, the procedure starts again from position j, etc.

To work with the Markov chain we need more detail about its analytic description. Let $X_i(t)$ denote the position of the chain at time $t \geq 0$ if it started at i at initial time $t = 0$, and let $P_{ij}(t)$ denote the probability of the transfer from i to j in time t, so that

$$\mathbf{P}(X_j(t) = j) = P_{ij}(t).$$

Then one can show that these *transition probabilities* $P(t) = \{P_{ij}(t)\}$ satisfy the following *Kolmogorov forward equations*:

$$\frac{d}{dt}P_{ij}(t) = \sum_{l=1}^{d} Q_{lj}P_{il}(t), \quad t \geq 0,$$

or in matrix form,

$$\frac{d}{dt}P(t) = P(t)Q.$$

On the other hand, the evolution of averages

$$T^t f(i) = \mathbf{E}f(X_i(t)) = \sum_{j=1}^{d} P_{ij}(t)f(j)$$

for any function f on the state space $\{1, ..., d\}$, i.e., for any vector $f \in \mathbf{R}^d$, satisfies the *Kolmogorov backward equations*

$$\frac{d}{dt}T^t f(i) = \sum_{j=1}^{d} Q_{ij}T^t f(j), \quad t \geq 0,$$

or in the vector form,

$$\frac{d}{dt}T^t f = QT^t f.$$

Consequently, T^t can be calculated as the following matrix exponent:

$$T^t f = e^{tQ} f.$$

Remark 5. *The values of the function f above will be denoted both by f_n and $f(n)$, reflecting two interpretations of f as a vector in $\mathbf{R}^d$ and a function on the state space.*

The matrix Q considered as a linear operator in $\mathbf{R}^d$ is called the *generator* of the Markov chain $X_i(t)$, and the operators T^t in $\mathbf{R}^d$ are called the *transition operators*. These operators form a semigroup, that is, they satisfy the equation $T^t T^s = T^{t+s}$ for all $s, t > 0$, which can be seen directly from the exponential representation $T^t = e^{tQ}$.

These definitions have natural extensions for the case of time-dependent matrices Q. Namely, let $\{Q(t)\} = \{(Q_{ij})(t)\}$ be a family of $d \times d$ square Q-matrices or Kolmogorov matrices depending piecewise continuously on time $t \geq 0$. The family $\{Q(t)\}$ specifies a (time inhomogeneous) *Markov chain* $X_{s,j}(t)$ on the state space $\{1, ..., d\}$, which is the following process. Starting from any time t and a current state i, one waits a $|Q_{ii}|(t)$-exponential random waiting time. After such time τ, the position jumps to a state j according to the distribution $(Q_{ij}/|Q_{ii}|)(t)$. Then at time $t + \tau$, the procedure starts again from position j, etc.

Let us denote by $X_{s,j}(t)$ the position of this process at time t if it was initiated at time s in the state i. The *transition probabilities* $P(s, t) = (P_{ij}(s, t))_{i,j=1}^{d}$, $s \le t$, defining the probabilities of migrating from i to j during the time segment $[s, t]$, are said to form the *transition matrix*, and the corresponding operators

$$U^{s,t} f(i) = \mathbf{E} f(X_{s,i}(t)) = \sum_{j=1}^{d} P_{ij}(s, t) f(j)$$

are called the *transition operators* of the Markov chain. The matrices $Q(t)$ define the time-dependent generator of the chain acting in $\mathbf{R}^d$ as

$$(Q(t)f)_n = \sum_{m \neq n} Q_{nm}(t)(f_m - f_n), \quad f = (f_1, \cdots, f_d).$$

The transition matrices satisfy the *Kolmogorov forward equation*

$$\frac{d}{dt} P_{ij}(s, t) = \sum_{l=1}^{d} Q_{lj}(t) P_{il}(s, t), \quad s \le t,$$

and the transition operators of this chain satisfy the *chain rule*, also called the *Chapman–Kolmogorov equation*, $U^{s,r} U^{r,t} = U^{s,t}$ for $s \le r \le t$, and the *Kolmogorov backward equations*

$$\frac{d}{ds}(U^{s,t} f)(i) = -\sum_{j=1}^{d} Q_{ij}(U^{s,t} f)(j), \quad s \le t.$$

A two-parameter family of operators $U^{s,t}$, $s \le t$, satisfying the chain rule is said to from a (backward) *propagator*.

Concerning discrete-time Markov chains $X_i(t)$ specified by the transition matrix $P = \{P_{ij}\}$ (briefly mentioned in Section 1.1), let us recall the evident fact that the probabilities P_{ij}^n of transitions $i \to j$ in time n (i.e., the probabilities of being in j at time t conditioned on the initial state i at time 0) form the matrix $P^n = \{P_{ij}^n\}$, which is the power of the transition matrix P: $P^n = (P)^n, n = 0, 1, \cdots$. The corresponding transition operators describing the dynamics of the averages act as multiplication by P^n, and hence will also be denoted by P^n:

$$P^n f(i) = \mathbf{E} f(X_i(n)) = (P^n f)(i) = \sum_{j} P_{ij}^n f(j).$$

A trivial but important modification to be mentioned is a similar setting but with time between jumps being any fixed time τ. This modification allows one to establish a close link between discrete- and continuous-time modeling. Namely, for a continuous-time Markov chain with transition operators T^t specified by the Q-matrix Q, let

$$\tau < (\max_i |Q_{ii}|)^{-1}.$$

Then one can define the discrete-time Markov chain on the same state space with the transition matrix $P^\tau = \{P_{ij}^\tau\}$, where

$$P_{ii}^\tau = 1 - \tau |Q_{ii}|, \quad P_{ij}^\tau = \tau Q_{ij}, \quad j \neq i. \tag{2.1}$$

In matrix form, this can be rewritten as

$$P^\tau = 1 + \tau Q,$$

where 1 denotes the unit matrix. Hence

$$(P^\tau)^n = (1 + \tau Q)^n.$$

Consequently, if $\tau \to 0$ and $n \to \infty$ in such a way that $n\tau \to t$, then

$$\lim (P^\tau)^n = e^{tQ} = T^t, \tag{2.2}$$

yielding an important link between the semigroup T^t and its discrete-time approximations P^τ.

Extension of these results to time-dependent chains (with P depending explicitly on t) is more or less straightforward. The transition probabilities just become time-dependent:

$$P_{ii}^{\tau,t} = 1 - \tau \sum_i |Q_{ii}(t)|, \quad P_{ij}^{\tau,t} = \tau Q_{ij}(t), \quad i \neq j, \tag{2.3}$$

or $P^{\tau,t} = 1 + \tau Q(t)$.

2.2 Preliminaries: ODEs and First-Order PDEs

Let us turn to the auxiliary results on ODEs and linear PDEs that we shall need. The results are mostly standard (see, e.g., [142, 192, 193]), and we will not prove them, but collect them in a form convenient for direct referencing.

Let $\dot{x} = g(x)$ be a (vector-valued) ordinary differential equation (ODE) in $\mathbf{R}^d$. If g is a Lipschitz function, then its solution $X_x(t)$ with the initial condition x is known to be well defined for all $t \in \mathbf{R}$. Hence one can define the lifting of this evolution on functions:

$$T^t f(x) = f(X_x(t)).$$

These *transition operators* act as continuous contraction operators on the space $C(\mathbf{R}^d)$ and in its subspace $C_\infty(\mathbf{R}^d)$, and they form a group, i.e., $T^t T^s = T^{s+t}$ for all s, t. The operator

$$Lf(x) = g(x) \cdot \frac{\partial f}{\partial x}$$

is called the *generator* of this group, because for all $f \in C_\infty(\mathbf{R}^d) \cap C^1(\mathbf{R}^d)$, one has

$$\frac{d}{dt}|_{t=0} T^t f = Lf.$$

If $g \in C^1(\mathbf{R}^d)$, then

$$\frac{d}{dt} T^t f = LT^t f = T^t Lf \tag{2.4}$$

for all t, so that the function $S(t, x) = T^t f(x) = f(X_x(t))$ satisfies the linear first-order partial differential equation (PDE)

$$\frac{\partial S}{\partial t} - g(x) \cdot \frac{\partial S}{\partial x} = 0, \tag{2.5}$$

with the initial condition $S(0, x) = f(x)$ for all $f \in C_\infty(\mathbf{R}^d) \cap C^1(\mathbf{R}^d)$. Solutions $X_x(t)$ to the ODE $\dot{x} = g(x)$ are referred to as the *characteristics* of the PDE (2.5), because the solutions to (2.5) are expressed in terms of $X_x(t)$.

Let us recall the basic result on the smooth dependence of the solutions $X_x(t)$ on the initial point x and the implied regularity of the solutions to (2.5) (key steps of the proof are provided in Section 4.1 in a more general case).

Proposition 2.2.1. *(i) Let $g \in C_{bLip}(\mathbf{R}^d)$. Then $X_x(t) \in C_{bLip}(\mathbf{R}^d)$ as a function of x and*

$$\|X_x(t)\|_{Lip} \le \exp\{t \|g\|_{Lip}\}. \tag{2.6}$$

(ii) Let $g \in C^1(\mathbf{R}^d)$. Then $X_x(t) \in C^1(\mathbf{R}^d)$ as a function of x and (recall that $\|g\|_{Lip} = \|g^{(1)}\|$)

$$\left\| \frac{\partial X_x(t)}{\partial x} \right\| = \sup_{j,x} \left\| \frac{\partial X_x(t)}{\partial x_j} \right\| \le \exp\{t \|g^{(1)}\|\} = \exp\{t \sup_{k,x} \left\| \frac{\partial g(x)}{\partial x_k} \right\| \}. \tag{2.7}$$

Moreover, if $f \in C^1(\mathbf{R}^d)$, then $T^t f \in C^1(\mathbf{R}^d)$ and

$$\|(T^t f)^{(1)}\| = \sup_{j,x} \left| \frac{\partial}{\partial x_j} f(X_x(t)) \right| \le \|f^{(1)}\| \left\| \frac{\partial X_x(t)}{\partial x} \right\| \le \|f^{(1)}\| \exp\{t \|g^{(1)}\|\}. \tag{2.8}$$

(iii) Let $g \in C^2(\mathbf{R}^d)$. Then $X_x(t) \in C^2(\mathbf{R}^d)$ as a function of x and

$$\left\|\frac{\partial^2 X_x(t)}{\partial x^2}\right\| = \sup_{j,i,x}\left\|\frac{\partial^2 X_x(t)}{\partial x_i \partial x_j}\right\| \le t\|g^{(2)}\|\exp\{3t\|g^{(1)}\|\}. \tag{2.9}$$

Moreover, if $f \in C^2(\mathbf{R}^d)$, then $T^t f \in C^2(\mathbf{R}^d)$ and

$$\|(T^t f)^{(2)}\| = \sup_{j,i,x}\left|\frac{\partial^2}{\partial x_j \partial x_i} f(X_x(t))\right|$$

$$\le \|f^{(2)}\|\exp\{2t\|g^{(1)}\|\} + t\|f^{(1)}\|\,\|g^{(2)}\|\exp\{3t\|g^{(1)}\|\}. \tag{2.10}$$

Let us also quote a result on the continuous dependence of the solutions on the r.h.s. and initial data.

Proposition 2.2.2. *Let $g_i \in C_{bLip}(\mathbf{R}^d)$, $i = 1, 2$, and let $X_x^i(t)$ denote the solutions of the equations $\dot{x} = g_i(x)$. Let $\|g_1 - g_2\| \le \delta$. Then*

$$|X_x^1(t) - X_y^2(t)| \le (t\delta + |x - y|)\exp\{t\|g_i\|_{Lip}\}, \quad i = 1, 2. \tag{2.11}$$

As follows from (2.4),

$$\frac{T^t f - f}{t} \to Lf, \quad t \to 0,$$

for all $f \in C_\infty(\mathbf{R}^d) \cap C^1(\mathbf{R}^d)$. As a corollary to previous results, let us obtain the rates of this convergence.

Proposition 2.2.3. *Let $g \in C^1(\mathbf{R}^d)$ and $f \in C_\infty(\mathbf{R}^d) \cap C^2(\mathbf{R}^d)$. Then*

$$\|T^t f - f\| \le t\|g\|\,\|f\|_{bLip}, \tag{2.12}$$

$$\left\|\frac{T^t f - f}{t} - Lf\right\| \le t\|g\|(\|g\|\,\|f^{(2)}\| + \|g\|_{Lip}\|f\|_{Lip}) \le t\|g\|\,\|g\|_{bLip}\|f\|_{C^2}. \tag{2.13}$$

Proof. By (2.4),

$$\|T^t f - f\| = \left\|\int_0^t T^s Lf\, ds\right\| \le t\|Lf\|,$$

implying (2.12). Next, by (2.4),

$$\frac{T^t f - f}{t} - Lf = \frac{1}{t}\int_0^t T^s Lf\, ds - Lf = \frac{1}{t}\int_0^t (T^s Lf - Lf)\, ds.$$

Applying (2.12) to the function Lf yields the first inequality in (2.13), the second inequality being a direct consequence. $\qquad\square$

A more general setting of nonautonomous equations $\dot{x} = g(t, x)$ is required, where g is Lipschitz in x and piecewise continuous in t. Here the solutions $X_{s,x}(t)$

with the initial point x at time s are well defined, and the transition operators form a two-parameter family:

$$U^{s,t} f(x) = f(X_{s,x}(t)). \tag{2.14}$$

These operators satisfy the chain rule $U^{s,r} U^{r,t} = U^{s,t}$, and the function $S(s, x) = U^{s,t} f(x)$ satisfies the PDE

$$\frac{\partial S}{\partial s} + g(s, x) \cdot \frac{\partial S}{\partial x} = 0, \tag{2.15}$$

with the initial (or terminal) condition $S(t, x) = f(x)$. As mentioned above, the operators $U^{s,t}$ satisfying the chain rule are said to form a *propagator*. Moreover, these operators form a *Feller propagator*, meaning that they also act in the space $C_\infty(\mathbf{R}^d)$ and depend strongly continuously on s and t. That is, if $f \in C_\infty(\mathbf{R}^d)$, then $U^{s,t} f(x)$ is a continuous function $t \mapsto U^{s,t} \in C_\infty(\mathbf{R}^d)$ for all s and a continuous function $s \mapsto U^{s,t} \in C_\infty(\mathbf{R}^d)$ for all t.

The solutions $X_{s,x}(t)$ to the ODE $\dot{x} = g(t, x)$ are called the *characteristics of the PDE* (2.15). The results of Propositions 2.2.1 and 2.2.2 (and their proofs) hold for the equation $\dot{x} = g(t, x)$ if all norms of functions of t, x are understood as the $\sup_t$ of their norms as functions of x, say

$$\|g\|_{Lip} = \sup_t \|g(t, .)\|_{Lip}, \quad \|g^{(2)}\| = \sup_t \|g^{(2)}(t, .)\|. \tag{2.16}$$

With this agreement, the following estimates hold:

$$\|X_{s,x}(t)\|_{Lip} \le \exp\{(t - s)\|g\|_{Lip}\}, \tag{2.17}$$

$$\|(U^{s,t} f)^{(1)}\| = \sup_{j,x} \left| \frac{\partial}{\partial x_j} f(X_{s,x}(t)) \right| \le \|f^{(1)}\| \exp\{(t - s)\|g^{(1)}\|\}, \tag{2.18}$$

$$\left\| \frac{\partial^2 X_{s,x}(t)}{\partial x^2} \right\| \le (t - s)\|g^{(2)}\| \exp\{3(t - s)\|g^{(1)}\|\}, \tag{2.19}$$

$$\|(U^{s,t} f)^{(2)}\| \le \|f^{(2)}\| \exp\{2(t - s)\|g^{(1)}\|\} + (t - s)\|f^{(1)}\| \|g^{(2)}\| \exp\{3(t - s)\|g^{(1)}\|\}, \tag{2.20}$$

with the derivatives on the l.h.s. existing whenever the derivatives on the r.h.s. exist.

As is easily seen, for propagators, the estimate (2.12) still holds, that is,

$$\|U^{s,t} f - f\| \le (t - s)\|g\| \|f\|_{bLip},$$

and in the estimate (2.13), an additional term appears on the r.h.s. depending on the continuity of g with respect to t. For instance, if g is Lipschitz in t with the constant

$$\varkappa = \sup_x \|g(., x)\|_{Lip},$$

then the analogue of the estimate (2.13) can be written as

$$\left\| \frac{U^{s,t} f - f}{t - s} - L_s f \right\| \leq (t - s)\|g\| \, \|g\|_{bLip} \|f\|_{C^2} + (t - s)\varkappa \|f\|_{bLip}. \quad (2.21)$$

Remark 6. *The evolution* $(s, x) \to X_{s,x}(t)$ *can be looked at as a deterministic Markov process with transition operators* $U^{s,t}$.

Finally, let us recall the result on the sensitivity of linear evolutions with respect to a parameter. Let us consider the following linear ODE in $\mathbf{R}^d$:

$$\dot{y} = A(t, x)y, \quad y \in \mathbf{R}^d, \quad t \in [0, T], \quad (2.22)$$

with the matrix A depending smoothly on a parameter $x = (x_1, \cdots, x_n) \in \mathbf{R}^n$. Here we shall work with y using their sup-norms. All function norms of A will be understood as the $\sup_{t \in [0,T]}$ of the corresponding norms as functions of x. For instance,

$$\|A\| = \sup_{t \in [0,T], x} \sup_i \sum_j |A_{ij}(t, x)|.$$

For the solution $y(t) = y(t, x)$ to (2.22) with some initial data $y_0(x)$ that may itself depend on x, we have the evident estimate

$$\|y(t)\|_{sup} \leq \exp\{t\|A\|\}\|y_0\|_{sup}.$$

Proposition 2.2.4. *Let* $A(t, .) \in C^2(\mathbf{R}^n)$. *Then* $y(t)$ *is also twice differentiable with respect to* x *and*

$$\left\| \frac{\partial y}{\partial x_i} \right\|_{sup} \leq \exp\{2t\|A\|\} \left(\left\| \frac{\partial y_0}{\partial x_i} \right\|_{sup} + t \left\| \frac{\partial A}{\partial x_i} \right\| \|y_0\|_{sup} \right), \quad (2.23)$$

$$\left\| \frac{\partial^2 y}{\partial x_i \partial x_j} \right\|_{sup} \leq \exp\{3t\|A\|\} \left(\left\| \frac{\partial^2 y_0}{\partial x_i \partial x_j} \right\|_{sup} + t \left\| \frac{\partial^2 A}{\partial x_i \partial x_j} \right\| \|y_0\|_{sup} \right.$$

$$\left. + 2t \sup_k \left\| \frac{\partial y_0}{\partial x_k} \right\|_{sup} \sup_k \left\| \frac{\partial A}{\partial x_k} \right\|_{sup} + 2t^2 \sup_k \left\| \frac{\partial A}{\partial x_k} \right\|^2 \|y_0\|_{sup} \right). \quad (2.24)$$

Proof. All these bounds arise from the solutions of the linear equations satisfied by the derivatives in question:

$$\frac{d}{dt} \frac{\partial y}{\partial x_i} = A(t, x) \frac{\partial y}{\partial x_i} + \frac{\partial A}{\partial x_i} y,$$

$$\frac{d}{dt}\frac{\partial^2 y}{\partial x_i \partial x_j} = A(t, x)\frac{\partial^2 y}{\partial x_i \partial x_j} + \frac{\partial A}{\partial x_j}\frac{\partial y}{\partial x_i} + \frac{\partial A}{\partial x_i}\frac{\partial y}{\partial x_j} + \frac{\partial^2 A}{\partial x_i \partial x_j}y.$$

For the justification (existence of the derivatives), we again refer to the textbooks on ODEs mentioned above.   □

2.3 Mean-Field Interacting Particle Systems

Let us turn to the basic setting of mean-field interacting particle systems with a finite number of types.

Let $\{Q(t, x)\} = \{(Q_{ij})(t, x)\}$ be a family of $d \times d$ square Q-matrices or Kolmogorov matrices depending Lipschitz continuously on a vector x from the closed simplex

$$\Sigma_d = \{x = (x_1, ..., x_d) \in \mathbf{R}_+^d : \sum_{j=1}^{d} x_j = 1\},$$

and piecewise continuously on time $t \geq 0$, so that $\|Q\|_{bLip} = \|Q\| + \|Q\|_{Lip}$, where we define

$$\|Q\| = \sup_{i,t,x} \sum_j |Q_{ij}(t, x)| < \infty, \quad \|Q\|_{Lip} = \sup_{x \neq y} \sup_{i,t} \frac{\sum_j |Q_{ij}(t, x) - Q_{ij}(t, y)|}{|x - y|}.$$
$$(2.25)$$

In view of the properties of Q-matrices, this implies

$$\|Q\| = 2\sup_{i,t,x} |Q_{ii}(t, x)| < \infty, \quad \|Q\|_{Lip} \leq 2\sup_{x \neq y} \sup_{i,t} \frac{\sum_{j \neq i} |Q_{ij}(t, x) - Q_{ij}(t, y)|}{|x - y|}.$$
$$(2.26)$$

The norms on the derivatives of the families of Kolmogorov matrices Q with respect to x can be, of course, introduced in various equivalent ways. For definiteness, we will use the following quantities, most convenient for our purposes:

$$\|Q\|_{C^1} = \|Q\| + \sup_i \sum_j \left|\sup_{k,t,x} \frac{\partial Q_{ij}}{\partial x_k}\right|, \quad \|Q\|_{C^2} = \|Q\|_{C^1} + \sup_i \sum_j \left|\sup_{k,l,t,x} \frac{\partial^2 Q_{ij}}{\partial x_k \partial x_l}\right|.$$
$$(2.27)$$

As above, this implies the estimates in terms of the transitions Q_{ij} with $j \neq i$:

$$\|Q\|_{C^1} \leq \|Q\| + 2\sup_i \sum_{j \neq i} \left|\sup_{k,t,x} \frac{\partial Q_{ij}}{\partial x_k}\right|, \quad \|Q\|_{C^2} \leq \|Q\|_{C^1} + 2\sup_i \sum_{j \neq i} \left|\sup_{k,l,t,x} \frac{\partial^2 Q_{ij}}{\partial x_k \partial x_l}\right|.$$
$$(2.28)$$

In what follows, the matrices Q will depend on an additional parameter controlled by the principal, but for the moment this dependence is not relevant and will be ignored.

Suppose we have a large number of particles distributed arbitrarily among the types $\{1, ..., d\}$. More precisely, our state space is $\mathbf{Z}_+^d$, the set of sequences of d nonnegative integers $n = (n_1, ..., n_d)$, where each n_i specifies the number of particles in the state i. Let N denote the total number of particles in state n: $N = n_1 + ... + n_d$. For $i \neq j$ and a state n with $n_i > 0$, denote by n^{ij} the state obtained from n by removing one particle of type i and adding a particle of type j, that is n_i, and n_j are changed to $n_i - 1$ and $n_j + 1$ respectively. The *mean-field interacting particle system* (in continuous time) specified by the family $\{Q\}$ is defined as the Markov chain on S with the generator

$$L_t^N f(n) = \sum_{i,j=1}^{d} n_i Q_{ij}(t, n/N)[f(n^{ij}) - f(n)]. \tag{2.29}$$

A probabilistic description of this process is as follows. Starting from any time and current state n, one attaches to each particle a $|Q_{ii}|(t, n/N)$-exponential random waiting time (where i is the type of this particle). If the shortest of the waiting times τ turns out to be attached to a particle of type i, this particle jumps to a state j according to the distribution $(Q_{ij}/|Q_{ii}|)(t, n/N)$. Briefly, with this distribution and at rate $|Q_{ii}|(t, n/N)$, any particle of type i can turn (migrate) into one of type j. After any such transition, the process starts again from the new state n^{ij}. Notice that since the number of particles N is preserved by any jump, this process is in fact a Markov chain with a finite state space.

Normalizing the states to $x = n/N \in \Sigma_d \cap \mathbf{Z}_+^d/N$ leads to the generator (also denoted by L_t^N, with some abuse of notation)

$$L_t^N f(n/N) = \sum_{i=1}^{d} \sum_{j=1}^{d} n_i Q_{ij}(t, n/N)[f(n^{ij}/N) - f(n/N)], \tag{2.30}$$

or equivalently,

$$L_t^N f(x) = \sum_{i=1}^{d} \sum_{j=1}^{d} x_i Q_{ij}(t, x) N[f(x - e_i/N + e_j/N) - f(x)], \quad x \in \mathbf{Z}_+^d/N, \tag{2.31}$$

where $e_1, ..., e_d$ denotes the standard basis in $\mathbf{R}^d$. Let us denote by $X^N(t) = X_{s,x}^N(t)$ the corresponding Markov chain and by $\mathbf{E} = \mathbf{E}_{s,x}$ the expectation with respect to this chain, where $x \in \Sigma_d \cap \mathbf{Z}_+^d/N$ denotes the initial state at time $s < t$. The transition operators of this chain will be denoted by $U_N^{s,t}$:

$$U_N^{s,t} f(x) = \mathbf{E} f(X_{s,x}^N(t)) = \mathbf{E}_{s,x} f(X^N(t)), \quad s \leq t. \tag{2.32}$$

Two versions of the notation in this formula (with (s, x) attached either to $\mathbf{E}$ or to $X^N(t)$) are both standard in the theory. Like any transition operators of a Markov chain, these operators satisfy the chain rule (or Chapman–Kolmogorov equation)

$$U_N^{s,r} U_N^{r,t} = U_N^{s,t}, \quad s \le r \le t,$$

and are said to form a propagator.

In the pressure and resistance setting of (1.10), the generator (2.31) reduces to

$$L_t^N f(x) = \sum_{i=1}^d \sum_{j=1}^d \varkappa x_i x_j (R_j(t, x) - R_i(t, x))^+ N[f(x - e_i/N + e_j/N) - f(x)].$$

(2.33)

Recall that the dual operator $(L_t^N)^*$ to the operator L_t^N is defined from the relation

$$\sum_{x=n/N \in \Sigma_d} [(L_t^N f)(x) g(x) - f(x)((L_t^N)^* g)(x)] = 0.$$

By a shift in the summation index, it is straightforward to see that the dual is given by the formula

$$(L_t^N)^* g(y) = \sum_{i=1}^d \sum_{j \ne i} [(y_i + 1/N) Q_{ij}(t, y + e_i/N - e_j/N) g(y + e_i/N - e_j/N)$$

$$- y_i Q_{ij}(t, y) g(y)],$$

(2.34)

where for convenience, it is set that $Q_{ij}(t, x) = 0$ for $x \notin \Sigma_d$. As is known from the theory of Markov chains, stable distributions g (also referred to as equilibrium probabilities) for the chains, defined by the generator $L_t^N = L^N$ in the time-homogeneous case, solve the equation $(L^N)^* g = 0$, and for large times t, these Markov chains converge to some of these equilibria.

In accordance with (2.1), a *mean-field interacting system in discrete time* related to the above-discussed mean-field interacting particle system in continuous time specified by the family $\{Q\}$ is defined as the Markov chain evolving in discrete time $t = k\tau, k \in \mathbf{N}$, with

$$\tau \le (N \| Q \|)^{-1},$$

(2.35)

with the transition probabilities

$$P_{nn^{ij}}^{\tau, t} = P_{nn^{ij}, N}^{\tau, t} = \tau n_i Q_{ij}(t, x), \quad i \ne j,$$

(2.36)

the probability of remaining in a given state n being $1 - \tau \sum_i n_i |Q_{ii}(t, x)|$. As follows from (2.36),

$$\frac{P_N^{\tau, t} f - f}{\tau} = L_t^N f$$

(2.37)

for all N, t, and τ. If Q does not depend explicitly on t, then the dependence on t disappears in (2.36), (2.37).

As one can expect and as will be shown, the theories of discrete- and continuous-time mean-field interacting systems are very similar. However, the discrete-time version is a bit more restrictive for the choice of parameters, since they are linked by (2.35).

Two asymptotic regimes for a mean-field interacting particle system (given by operators L_t^N) are studied in the first part of this book: $N \to \infty$ and $t \to \infty$. In full detail, the limit $\lim_{t\to\infty} \lim_{N\to\infty}$ will be studied. An important question is whether the order of applying these limits can be interchanged. In other words, what can be said about the $t \to \infty$ limit of a chain with a finite number of particles N (which is described essentially by the highly multidimensional equation $(L^N)^* g = 0$) if we know the large-time behavior of the simpler limiting dynamics $N \to \infty$ (given by (2.40) below).

2.4 Dynamic LLN: Smooth Coefficients

As above, the functional norms of functions of two variables (t, x) with $x \in \Sigma_d$ will mean the $\sup_t$ of their respective norms as functions of x.

Our main interest concerns the asymptotic behavior of these chains as $N \to \infty$. To this end, let us observe that for $f \in C^1(\Sigma_d)$,

$$\lim_{N\to\infty, n/N\to x} N[f(n^{ij}/N) - f(n/N)] = \frac{\partial f}{\partial x_j}(x) - \frac{\partial f}{\partial x_i}(x),$$

so that

$$\lim_{N\to\infty, n/N\to x} L_t^N f(n/N) = \Lambda_t f(x),$$

where

$$\Lambda_t f(x) = \sum_{i=1}^{d}\sum_{j\neq i} x_i Q_{ij}(t, x)[\frac{\partial f}{\partial x_j} - \frac{\partial f}{\partial x_i}](x)$$

$$= \sum_{k=1}^{d}\sum_{i\neq k}[x_i Q_{ik}(t, x) - x_k Q_{ki}(t, x)]\frac{\partial f}{\partial x_k}(x). \qquad (2.38)$$

More precisely, if $f \in C^2(\Sigma_d)$, then by Taylor's formula,

$$L_t^N f(x) - \Lambda_t f(x) = \frac{1}{2N}\sum_{i=1}^{d}\sum_{j\neq i} x_i Q_{ij}(t, x)[\frac{\partial^2 f}{\partial x_i^2} + \frac{\partial^2 f}{\partial x_j^2} - \frac{\partial^2 f}{\partial x_i \partial x_j}](\theta),$$

for $x = n/N$, with some $\theta \in \Sigma_d$, and thus

$$\|L_t^N f - \Lambda_t f\| \leq \frac{1}{N}\|f^{(2)}\|\,\|Q\|. \tag{2.39}$$

The limiting operator $\Lambda_t f$ is a first-order PDO with characteristics solving the equations

$$\dot{x}_k = \sum_{i \neq k}[x_i Q_{ik}(t, x) - x_k Q_{ki}(t, x)] = \sum_{i=1}^{d} x_i Q_{ik}(t, x), \quad k = 1, ..., d, \tag{2.40}$$

called the *kinetic equations* for the process of interaction described above. In vector form, this system can be rewritten as

$$\dot{x} = Q^T(t, x)x = xQ(t, x), \tag{2.41}$$

where Q^T is the transpose matrix to Q. (In the second notation $xQ(t, x)$, the vector x is understood as a row vector allowing for multiplication by Q from the right.)

The corresponding transition operators act on $C(\Sigma_d)$ as

$$U^{s,t} f(x) = f(X_{s,x}(t)), \quad s \leq t. \tag{2.42}$$

For the case of the operator (2.33), the limiting operator takes the form

$$\Lambda_t f(x) = \sum_{i,j=1}^{d} \varkappa x_i x_j [R_j(t, x) - R_i(t, x)]^+ \left[\frac{\partial f}{\partial x_j} - \frac{\partial f}{\partial x_j}\right]$$

$$= \sum_{i,j=1}^{d} \varkappa x_i x_j [R_j(t, x) - R_i(t, x)]\frac{\partial f}{\partial x_j}, \tag{2.43}$$

and the characteristics (or the kinetic equations) become

$$\dot{x}_j = \sum_{i=1}^{d} \varkappa x_i x_j [R_j(t, x) - R_i(t, x)]. \tag{2.44}$$

For the sake of clarity, we often derive results first for time-independent Q and then comment on the (usually straightforward) modifications required for the general case (time-dependent versions are important only for the second part of the book).

Thus we shall denote by $X_x(t)$ the characteristics and by $X_x^N(t)$ the Markov chain starting at x at time $t = 0$. The generators of these processes also become time-homogeneous: $L_t^N = L^N$, $\Lambda_t = \Lambda$. The corresponding transition operators $U^{s,t}$ depend only on the difference $t - s$, and the operators $U^t = U^{0,t}$, defined as

$U^t f(x) = f(X_x(t))$, form a semigroup. The transition operators for the Markov chain $X_x^N(t)$ are

$$U_N^t f(x) = \mathbf{E} f(X_x^N(t)) = \mathbf{E}_x f(X^N(t)).$$

Let us write down explicitly the straightforward estimates of the norms of the r.h.s. $g(t, x) = x Q(t, x)$ of (2.41), as a function of x, in terms of the norms of Q introduced in (2.26) and (2.28):

$$\|g\| \leq \|Q\|, \quad \|g\|_{Lip} \leq \|Q\|_{bLip}, \quad \|g\|_{bLip} \leq 2\|Q\|_{bLip}, \quad \|g^{(2)}\| \leq 2\|Q\|_{C^2},$$
$$(2.45)$$

because

$$\frac{\partial g_k}{\partial x_l} = Q_{lk} + \sum_j x_j \frac{\partial Q_{jk}}{\partial x_l}, \quad \frac{\partial^2 g_k}{\partial x_l \partial x_m} = \frac{\partial Q_{lk}}{\partial x_m} + \frac{\partial Q_{mk}}{\partial x_l} + \sum_j x_j \frac{\partial^2 Q_{jk}}{\partial x_l \partial x_m}.$$

Under (2.26), system (2.41) is easily seen to be well posed in Σ_d (see the remark below for additional comments and the proof in the more general infinite-dimensional setting in Section 4.1), that is, for all $x \in \Sigma_d$, the solution $X_x(t)$ to (2.41) with Q not depending on t and with the initial condition x at time $t = 0$ is well defined and belongs to Σ_d for all times $t > 0$ (not necessarily for $t < 0$).

Remark 7. *(i) Unlike the general setting (2.14), where the condition $t > 0$ was not essential, here it is essential, because the preservation of the simplex Σ_d holds only in forward time. It holds because $\dot{x}_k \geq 0$ whenever $x_k = 0$ and $x \in \Sigma_d$, which does not allow a trajectory to cross the boundary of Σ_d. Hence the operators Φ^t are defined as operators in $C(\Sigma_d)$ only for $t > 0$. (ii) It is seen from the structure of (2.41) that if $x_k \neq 0$, then $(X_x(t))_k \neq 0$ for all $t \geq 0$. Hence the boundary of Σ_d is not attainable for this semigroup, but depending on Q, it can be gluing or not. For instance, if all elements of Q never vanish, then the points $X_x(t)$ never belong to the boundary of Σ_d for $t > 0$, even if the initial point x does. In fact, in this case, $\dot{x}_k > 0$ whenever $x_k = 0$ and $x \in \Sigma_d$.*

Our first objective now is to show that the Markov chains $X_x^N(t)$ do in fact converge to the deterministic evolution $X_x(t)$ in the sense that the corresponding transition operators converge (the so-called weak convergence of Markov processes), the previous arguments showing only that their generators converge to sufficiently smooth functions. We are also interested in precise rates of convergence.

The next elementary result concerns the (unrealistic) situation with optimal regularity of all objects concerned.

Theorem 2.4.1. *Let all the elements $Q_{ij}(x)$ belong to $C^2(\Sigma_d)$ and $f \in C^2(\Sigma_d)$. Then*

$$\sup_{x \in \mathbf{Z}_+^d/N} |U_N^t f(x) - U^t f(x)| \leq \frac{t\|Q\|}{N} (\|f^{(2)}\| + 2t\|f\|_{bLip} \|Q\|_{C^2}) \exp\{3t\|Q\|_{bLip}\},$$
$$(2.46)$$

and for all x and n/N,

$$|U_N^t f(n/N) - U^t f(x)|$$

$$\leq \left[\frac{t\|Q\|}{N} (\|f^{(2)}\| + 2t\|f\|_{bLip}\|Q\|_{C^2}) + \|f\|_{bLip}\|x - n/N\| \right] \exp\{3t\|Q\|_{bLip}\}.$$
(2.47)

Finally, if the initial states n/N converge to a point $x \in \Sigma_d$ as $N \to \infty$, then

$$\sup_{0 \leq t \leq T} |U_N^t f(n/N) - U^t f(x)| \to 0, \quad N \to \infty,$$
(2.48)

for all T and all $f \in C(\Sigma_d)$.

Proof. To compare the semigroups, we shall use the following standard trick. We write

$$(U^t - U_N^t)f = U_N^{t-s}U^s|_{s=0}^t f = \int_0^t \frac{d}{ds}U_N^{t-s}U^s f \, ds = \int_0^t U_N^{t-s}(\Lambda - L^N)U^s f \, ds.$$
(2.49)

Let us apply this equation to an $f \in C^2(\Sigma_d)$. By (2.10) and the second estimate of (2.45),

$$\|(U^t f)^{(2)}\| \leq (\|f^{(2)}\| + 2t\|f\|_{bLip}\|Q\|_{C^2}) \exp\{3t\|Q\|_{bLip}\}.$$

Hence by (2.39),

$$\|(L^N - \Lambda)U^s f\| \leq \frac{\|Q\|}{N}(\|f^{(2)}\| + 2t\|f\|_{bLip}\|Q\|_{C^2}) \exp\{3t\|Q\|_{bLip}\}.$$

Consequently, (2.46) follows from (2.49) and the contraction property of U_N^t.

Equation (2.47) follows, because by (2.8) and (2.45),

$$|U^t f(x) - U^t f(y)| \leq \|U^t f\|_{Lip}\|x - y\| \leq \exp\{t\|Q\|_{bLip}\}\|f\|_{Lip}\|x - y\|.$$

The last statement is obtained because of the possibility to approximate any function $f \in C(\Sigma_d)$ by smooth functions. □

By the definition of the propagators U^t, U_N^t, equations (2.46)–(2.48) can be written in terms of the averages of the Markov chains X_x^N. For instance, (2.48) takes the form

$$\sup_{0 \leq t \leq T} |\mathbf{E}f(X_{n/N}^N(t)) - f(X_x(t))| \to 0, \quad N \to \infty.$$
(2.50)

For time-dependent Q, everything remains the same:

Theorem 2.4.2. *Let all the elements $Q_{ij}(t, x)$ belong to $C^2(\Sigma_d)$ as functions of x and suppose they are piecewise continuous as functions of t. Let $f \in C^2(\Sigma_d)$. Then*

$$\sup_{x \in \mathbf{Z}_+^d/N} |U_N^{s,t} f(x) - U^{s,t} f(x)|$$

$$\leq \frac{(t-s)\|Q\|}{N}(\|f^{(2)}\| + 2(t-s)\|f\|_{bLip}\|Q\|_{C^2})\exp\{3(t-s)\|Q\|_{bLip}\}, \tag{2.51}$$

and the corresponding analogue of (2.47) holds.

The proof is also the same, though instead of (2.49), one uses its version for propagators:

$$(U^{s,t} - U_N^{s,t})f = -U_N^{s,r} U^{r,t}|_{r=s}^t f = \int_s^t U_N^{s,r}(\Lambda_r - L_r^N)U^{r,t} f \, dr. \tag{2.52}$$

The LLN for the discrete-time setting is also analogous. Namely, the following result holds, where we have returned to the time-homogeneous setting.

Theorem 2.4.3. *Let all the elements* $Q_{ij}(x)$ *belong to* $C^2(\Sigma_d)$ *and* $f \in C^2(\Sigma_d)$. *Let a discrete-time Markov chain be defined by* (2.35)–(2.37). *Then*

$$\sup_{n \in \mathbf{Z}_+^d} |(P_N^\tau)^k f(n/N) - U^t f(n/N)|$$

$$\leq \frac{2t\|Q\|}{N}(\|f^{(2)}\| + 2t\|f\|_{bLip}\|Q\|_{C^2})\exp\{3t\|Q\|_{bLip}\}, \tag{2.53}$$

for $t = k\tau$, *and for all* x *and* n/N,

$$|(P_N^\tau)^k f(n/N) - U^t f(x)|$$

$$\leq \left[\frac{2t\|Q\|}{N}(\|f^{(2)}\| + 2t\|f\|_{bLip}\|Q\|_{C^2}) + \|f\|_{bLip}\|x - n/N\|\right]\exp\{3t\|Q\|_{bLip}\}. \tag{2.54}$$

And of course, the analogue of convergence (2.48) *also holds.*

Proof. We have

$$((P_N^\tau)^k - U^{\tau k})f = (P_N^\tau)^{(k-1)}(P_N^\tau - U^\tau)$$
$$+ (P_N^\tau)^{(k-2)}(P_N^\tau - U^\tau)U^\tau + \cdots + (P_N^\tau - U^\tau)U^{\tau(k-1)}.$$

By (2.39), (2.37), and (2.13),

$$\|(P_N^\tau - U^\tau)f\| = \tau \left\|\frac{P_N^\tau f - f}{\tau} - \frac{U^\tau f - f}{\tau}\right\|$$

$$\leq \tau \left(\frac{\|Q\|}{N}\|f^{(2)}\| + \tau\|Q\|^2\|f^{(2)}\| + \tau\|Q\|\|Q\|_{bLip}\|f\|_{bLip}\right)$$

$$\leq \frac{\tau}{N}(2\|Q\|\,\|f^{(2)}\| + \|Q\|_{bLip}\|f\|_{bLip}),$$

where (2.35) was taken into account. Consequently, taking into account (2.10), we obtain

$$\|(P_N^{\tau})^k f - U^{\tau k} f\| \leq \exp\{3t\|Q\|_{bLip}\}$$

$$\times \left(\frac{2\tau n}{N}\|Q\|\,\|f^{(2)}\| + \frac{\tau}{N}\sum_{k=0}^{n-1}(\|Q\|_{bLip}\|f\|_{bLip} + 2k\tau\|Q\|_{C^2}\|f\|_{bLip})\right),$$

yielding (2.53). The estimate (2.54) follows as in Theorem 2.4.1.   □

2.5 Dynamic LLN: Lipschitz Coefficients

Let us move now to a more realistic situation in which Q is assumed to be only Lipschitz, that is, (2.26) holds.

Theorem 2.5.1. *Let the functions $Q(t, x)$ be piecewise continuous in t and belong to C_{bLip} as functions of x with $\|Q\|_{bLip} = \sup_t \|Q(t,.)\| \leq \omega$ with some ω. Suppose the initial data $x(N) = n/N$ of the Markov chains $X_{s,x(N)}^N(t)$ converge to a certain x in $\mathbf{R}^d$ as $N \to \infty$. Then these Markov chains converge to the deterministic evolution $X_{s,x}(t)$, in the weak sense:*

$$|\mathbf{E}f(X_{s,x(N)}^N(t)) - f(X_{s,x}(t))| \to 0, \ \text{as } N \to \infty, \tag{2.55}$$

or in terms of the transition operators

$$|U_N^{s,t}f(x) - U^{s,t}f(x)| \to 0, \ \text{as } N \to \infty, \tag{2.56}$$

for all $f \in C(\Sigma_d)$, the convergence being uniform in x whenever the convergence $x(N) \to x$ is uniform.

For smooth or Lipschitz f, the following rates of convergence are valid:

$$|\mathbf{E}f(X_{s,x(N)}^N(t)) - f(X_{s,x(N)}(t))| \leq C(t-s)\exp\{3(t-s)\|Q\|_{bLip}\}$$
$$\left(\frac{(t-s)^{1/2}}{N^{1/2}}\|Q\|_{bLip}(d + \|Q\|)\|f\|_{bLip} + \frac{\|Q\|}{N}\|f^{(2)}\|\right), \tag{2.57}$$

$$|\mathbf{E}f(X_{s,x(N)}^N(t)) - f(X_{s,x(N)}(t))|$$

$$\leq C\exp\{3(t-s)\|Q\|_{bLip}\}(d + \|Q\|)\|Q\|_{bLip}\frac{(t-s)^{1/2}}{N^{1/2}}\|f\|_{bLip}, \tag{2.58}$$

$$|f(X_{s,x(N)}(t)) - f(X_{s,x}(t))| \leq \exp\{(t-s)\|Q\|_{bLip}\}\|f\|_{bLip}\|x(N) - x\|, \quad (2.59)$$

with a constant C.

Remark 8. *The dependence on t and d is not essential here, but the latter becomes crucial for dealing with infinite state spaces, while the former is crucial for dealing with a forward-looking principal.*

Remark 9. *Assuming the intermediate regularity of Q, that is, assuming $Q \in C^1(\Sigma_d)$ with all first-order derivatives being Hölder continuous with a fixed index $\alpha \in (0,1)$, will yield intermediate rates of convergence between $1/N$ and $1/\sqrt{N}$ above.*

Proof. To shorten the formulas, let us write down a proof for time-independent Q.

The Lipschitz continuity (2.59) of the solutions is a consequence of Proposition 2.2.1. Next, since any function $f \in C(\mathbf{R}^d)$ can be approximated by functions from $C^2(\mathbf{R}^d)$, the convergence (2.55) follows from (2.57) and (2.59). Thus it remains to show (2.57) and (2.58).

The main idea is to approximate all Lipschitz continuous functions involved by the smooth ones. Namely, choosing an arbitrary mollifier χ (nonnegative infinitely smooth even function on $\mathbf{R}$ with compact support and $\int \chi(w)\,dw = 1$) and the corresponding mollifier $\phi(y) = \prod \chi(y_j)$ on $\mathbf{R}^{d-1}$, let us define, for every function V on Σ_d, its approximation

$$\Phi_\delta[V](x) = \int_{R^{d-1}} \frac{1}{\delta^{d-1}} \phi\left(\frac{y}{\delta}\right) V(x-y)\,dy = \int_{R^{d-1}} \frac{1}{\delta^{d-1}} \phi\left(\frac{x-y}{\delta}\right) V(y)\,dy.$$

Notice that Σ_d is a $(d-1)$-dimensional object, so that every V on it can be considered a function of the first $(d-1)$ coordinates of a vector $x \in \Sigma_d$ (continued to $\mathbf{R}^{d-1}$ in an arbitrarily continuous way). It follows that

$$\|\Phi_\delta[V]\|_{C^1} = |\Phi_\delta[V]\|_{bLip} \leq \|V\|_{bLip} \quad (2.60)$$

for all δ and

$$|\Phi_\delta[V](x) - V(x)| \leq \int \frac{1}{\delta^{d-1}} \phi\left(\frac{y}{\delta}\right) |V(x-y) - V(x)|\,dy$$

$$\leq \|V\|_{Lip} \int_{R^{d-1}} \frac{1}{\delta^{d-1}} \phi\left(\frac{y}{\delta}\right) |y|_1\,dy \leq \delta(d-1)\|V\|_{Lip} \int_R |w|\chi(w)\,dw. \quad (2.61)$$

Remark 10. *We care about the dimension d in the estimates only for future use (here it is irrelevant). By a different choice of mollifier ϕ one can get rid of d in (2.61), but then it would pop up in (2.62), which is avoided with our ϕ.*

Next, the norm $\|\Phi_\delta[V]\|_{C^2}$ does not exceed the sum of the norm $\|\Phi_\delta[V]\|_{C^1}$ and the supremum of the Lipschitz constants of the functions

$$\frac{\partial}{\partial x_j}\Phi_\delta[V](x) = \int \frac{1}{\delta^d}\left(\frac{\partial}{\partial x_j}\phi\right)\left(\frac{y}{\delta}\right)V(x-y)\,dy.$$

Hence

$$\|\Phi_\delta[V]^{(2)}\| \le \|V\|_{bLip}\frac{1}{\delta}\int |\chi'(w)|\,dw,$$

$$\|\Phi_\delta[V]\|_{C^2} \le \|V\|_{bLip}\left(1+\frac{1}{\delta}\int |\chi'(w)|\,dw\right). \tag{2.62}$$

Let $U_{N,\delta}^t$ and U_δ^t denote the same transition operators as above but constructed with respect to the matrices

$$\Phi_\delta[Q](x) = \int \frac{1}{\delta^d}\phi\left(\frac{y}{\delta}\right)Q(x-y)\,dy$$

rather than Q. Notice that $\Phi_\delta[Q](x)$ are also Q-matrices for every δ.

Similarly, we denote by $L^{N,\delta}$ and Λ^δ the corresponding generators and by $X_x^\delta(t)$ the characteristics with $\Phi_\delta[Q]$ used instead of Q.

By (2.61) and (2.11),

$$\|X_t(x) - X_t^\delta(x)\| \le C\delta td\|Q\|_{bLip}\exp\{t\|Q\|_{bLip}\},$$

and hence

$$|U^t f(x) - U_\delta^t f(x)| = |f(X_t(x)) - f(X_t^\delta(x))|$$
$$\le C\|Q\|_{bLip}\|f\|_{bLip}\delta td\,\exp\{t\|Q\|_{bLip}\}. \tag{2.63}$$

Moreover, since

$$\|(L^{N,\delta} - L^N)f\| \le \delta(d-1)\|f\|_{bLip}\|Q\|_{bLip}\int_{\mathbf{R}} |w|\chi(w)\,dw,$$

it follows by (2.49) applied to the propagators U_N and $U_{N,\delta}$ that the same estimate (2.63) holds for the difference $U_{N,\delta}^t - U_N^t$:

$$\|U_{N,\delta}^t f - U_N^t f\| \le C\|Q\|_{bLip}\|f\|_{bLip}\delta td\,\exp\{t\|Q\|_{bLip}\}. \tag{2.64}$$

By (2.46) and (2.62),

$$\|U_{N,\delta}^t f - U_\delta^t f\| \le \frac{t\|Q\|}{N}(\|f^{(2)}\| + \frac{Ct}{\delta}\|Q\|_{bLip}\|f\|_{bLip}). \tag{2.65}$$

Therefore,

$$\|U_N^t f - U^t f\| \le \|U_N^t f - U_{N,\delta}^t f\| + \|U_{N,\delta}^t f - U_\delta^t f\| + \|U_\delta^t f - U^t f\|$$

$$\leq Ct \left(\delta d \|Q\|_{bLip} \|f\|_{bLip} + \frac{\|Q\|}{N} \|f^{(2)}\| + \frac{t}{N\delta} \|Q\|_{bLip} \|Q\| \|f\|_{bLip} \right) \exp\{3t\|Q\|_{bLip}\}.$$

Thus choosing $\delta = \sqrt{t/N}$ makes the decay rate of δ and $t/(N\delta)$ equal, yielding (2.57).

Finally, if f is only Lipschitz, we approximate it by $\tilde{f} = \Phi_{\tilde{\delta}}[f]$, so that the second derivative of $\Phi_{\tilde{\delta}}[f]$ is bounded by $\|f\|_{bLip}/\tilde{\delta}$. By the contraction property of U_N^t and U^t,

$$\|U_N^t(f - \tilde{f})\| \leq \|f - \tilde{f}\| \leq Cd\tilde{\delta}\|f\|_{bLip},$$
$$\|U^t(f - \tilde{f})\| \leq \|f - \tilde{f}\| \leq Cd\tilde{\delta}\|f\|_{bLip}.$$

Thus the rates of convergence for f become of order

$$[d\tilde{\delta} + t\delta d\|Q\|_{bLip} + \frac{t^2}{N\delta}\|Q\|_{bLip}\|Q\| + \frac{t}{N\tilde{\delta}}]\|f\|_{bLip} \exp\{3t\|Q\|_{bLip}\}.$$

Choosing $\delta = \tilde{\delta} = \sqrt{t/N}$ yields (2.58). $\qquad\square$

The corresponding result for the discrete-time setting is again fully analogous.

Theorem 2.5.2. *Let the functions $Q(x)$ belong to C_{bLip}. Let the discrete-time Markov chains $X_x^{N,\tau}(t)$ be defined by (2.35)–(2.37). Suppose the initial data $x(N) = n/N$ of these Markov chains converge to a certain x in $\mathbf{R}^d$ as $N \to \infty$ (and thus $\tau \to 0$ by (2.35)). Then these Markov chains converge to the deterministic evolution $X_x(t)$, in the weak sense:*

$$|\mathbf{E}f(X_{x(N))}^{N,\tau}(t)) - f(X_x(t))| \to 0, \quad as \ N \to \infty, \tag{2.66}$$

for all $f \in C(\Sigma_d)$, the convergence being uniform in x whenever the convergence $x(N) \to x$ is uniform.

For smooth or Lipschitz f, the following rates of convergence are valid:

$$|\mathbf{E}f(X_{x(N)}^{N,\tau}(t)) - f(X_{x(N)}(t))|$$

$$\leq Ct \exp\{3t\|Q\|_{bLip}\} \left(\frac{t^{1/2}}{N^{1/2}}(d + \|Q\|_{bLip}^2)\|f\|_{bLip} + \frac{\|Q\|}{N}\|f^{(2)}\| \right), \tag{2.67}$$

$$|\mathbf{E}f(X_{x(N)}^{N,\tau}(t)) - f(X_{x(N)}(t))| \leq C \exp\{3t\|Q\|_{bLip}\}(d + \|Q\|_{bLip}^2) \frac{t^{1/2}}{N^{1/2}}\|f\|_{bLip}, \tag{2.68}$$

with a constant C.

Proof. The only modification as compared with the proof of Theorem 2.5.2 is the necessity to estimate $(P_{N,\delta}^\tau)^k - (P_N^\tau)^k$ instead of $U_{N,\delta}^t - U_N^t$. This is done as in the proof of Theorem 2.4.3 using the identity

$$((P_N^\tau)^k - (P_{N,\delta}^\tau)^k)f = (P_N^\tau)^{(n-1)}(P_N^\tau - P_{N,\delta}^\tau)$$

$$+(P_N^\tau)^{(n-2)}(P_N^\tau - P_{N,\delta}^\tau)P_{N,\delta}^\tau + \cdots + (P_N^\tau - P_{N,\delta}^\tau)(P_{N,\delta}^\tau)^{k-1}.$$

$\square$

2.6 Dynamic LLN with Major Players

As discussed in Section 1.4, we are mostly interested in the presence of a principal that may exert pressure on small players on the level described by the parameter b from a bounded convex subset of a Euclidean space. A mean-field interacting particle system controlled by the principal will be generated by (2.29) with the coefficients depending on the control parameter b of the principal (and not depending on time, for simplicity):

$$L^{N,b} f(n) = \sum_{i,j=1}^d n_i Q_{ij}(n/N, b)[f(n^{ij}) - f(n)]. \tag{2.69}$$

In the simplest setting, which we refer to as the "best-response principle," one can imagine the principal choosing the value of b^* maximizing some current profit $B(x, b, N)$ for given x, N, that is, via (1.5):

$$b^*(x, N) = arg max\, B(x, ., N). \tag{2.70}$$

If there exists a limit $b^*(x) = \lim b^*(x, N)$, then the limiting evolution (1.3) turns into the evolution (1.6):

$$\dot{x}_k = \sum_{i=1}^d x_i Q_{ik}(x, b^*(x)), \quad k = 1, ..., d, \tag{2.71}$$

or in particular in the pressure and resistance framework, into the evolution (1.11):

$$\dot{x}_j = \sum_i \varkappa x_i x_j [R_j(x, b^*(x)) - R_i(x, b^*(x))], \quad j = 1, ..., d. \tag{2.72}$$

Remark 11. *A more strategically thinking principal will be discussed in Chapter 3.*

The corresponding modification of Theorem 2.5.1 is straightforward, yielding the following result (using time-homogeneous Q for simplicity).

Theorem 2.6.1. *Assume*

$$|b^*(x, N) - b^*(x)| \leq \epsilon(N), \tag{2.73}$$

with some $\epsilon(N) \to 0$, as $N \to \infty$ and some function $b^*(x)$, and let the functions $Q(x, b)$ (or in particular, $R_j(x, b)$ in the pressure and resistance framework), $b^*(x, N)$, $b^*(x)$ belong to C_{bLip} as functions of their variables with norms uniformly bounded by some ω. Suppose the initial data $x(N) = n/N$ of the Markov chains $X^N_{x(N)}(t)$ converge to a certain x in $\mathbf{R}^d$ as $N \to \infty$. Then these Markov chains converge to the deterministic evolution $X_x(t)$ solving (2.71) (or (2.72) respectively):

$$|\mathbf{E} f(X^N_{x(N))}(t) - f(X_x(t))| \to 0, \;\; as \; N \to \infty, \tag{2.74}$$

for all $f \in C(\Sigma_d)$, the convergence being uniform in x whenever the convergence $x(N) \to x$ is uniform. For Lipschitz f, the estimate (2.59) holds, and (2.58) generalizes to

$$|\mathbf{E} f(X^N_{x(N)}(t) - f(X_{x(N)}(t))| \leq C(\omega, t) \left(\frac{dt^{1/2}}{N^{1/2}} + t\epsilon(N) \right) \|f\|_{bLip}. \tag{2.75}$$

Similarly, there can be several, say K, major players or principals that may exert pressure on small players. Assume that their control parameters are b_j (each chosen from some bounded closed domain in a Euclidean space) and their profits are given by some Lipschitz continuous functions $B_j(x, b_1, \cdots, b_K, N)$ depending on the distribution x of small players. Recall that a *Nash equilibrium* in a game of K players given by these payoffs (for any fixed x, N) is a profile of "no regret" strategies $(b_1^*(x, N), \cdots, b_K^*(x, N))$, that is, they are such that unilateral deviation from such strategies cannot be profitable:

$$B_j(x, b_1^*(x, N), \cdots, b_K^*(x, N), N)$$

$$= \max_{b_j} B_j(x, b_1^*(x, N), \cdots, b_{j-1}^*(x, N), b_j, b_{j+1}^*(x, N), \cdots, b_K^*(x, N), N) \tag{2.76}$$

for all j.

Assume that for $x \in \Sigma_d$ there exists a branch of such Nash equilibria $(b_1^*(x, N), \cdots, b_K^*(x, N))$ depending continuously on x such that there exists a limit

$$(b_1^*(x), \cdots, b_K^*(x)) = \lim_{N \to \infty} (b_1^*(x, N), \cdots, b_K^*(x, N)). \tag{2.77}$$

The dynamics (2.71) or (2.72) and Theorem 2.6.1 extend automatically to the case of major players adhering to these local (x-dependent) Nash equilibria. For instance, (2.72) takes the form

$$\dot{x}_j = \sum_i \varkappa x_i x_j [R_j(x, b_1^*(x), \cdots, b_K^*(x)) - R_i(x, b_1^*(x), \cdots, b_K^*(x))], \quad j = 1, ..., d.$$

(2.78)

2.7 Dynamic LLN with Distinguished (Tagged) Player

Looking for the behavior of some particular distinguished or tagged particle inside the pool of a large number of indistinguishable ones is a well-known useful tool in statistical mechanics and experimental biology. Let us extend here the results of Sections 2.4 and 2.5 to the case of a tagged agent following different transition rules from those of the crowd. This analysis will be used in Section 5.3.

Under the setting of Theorem 2.5.1, let us assume that one distinguished agent in the group of N players deviates from the general rules, moving according to the transition Q-matrix $Q^{dev}(t, x)$. Then the natural state space for such a Markov chain will be $\{1, \cdots, d\} \times \Sigma_d$, the first coordinate j denoting the position of the tagged player. Instead of (2.31), the generator of this Markov chain becomes

$$L_t^{N,dev} f(j, x) = \sum_k Q_{jk}^{dev}(t, x)(f(k, x) - f(j, x))$$

$$+ \sum_i (x_i - \delta_i^j/N) \sum_{k \neq i} Q_{ik}(t, x) [f(j, x - e_i/N + e_k/N) - f(j, x)]. \quad (2.79)$$

Let $U_N^{s,t}$ denote the transition operators of this Markov chain.
For smooth f and as $N \to \infty$, the operators (2.79) converge to the operator

$$\Lambda_t^{dev} f(j, x) = \sum_k Q_{jk}^{dev}(t, x)(f(k, x) - f(j, x))$$

$$+ \sum_i x_i \sum_{k \neq i} Q_{ik}(t, x) \left[\frac{\partial f}{\partial x_k} - \frac{\partial f}{\partial x_i} \right] (j, x), \quad (2.80)$$

with the rates of convergence

$$\|L_t^{N,dev} f - \Lambda_t^{dev} f\| = \sup_{j,x} |(L_t^{N,dev} - \Lambda_t^{dev}) f(j, x)| \leq \frac{\|Q\|}{N} (\|f^{(2)}\| + 2\|f\|),$$

(2.81)

where

$$\|f^{(2)}\| = \sup_{j,i,k,x} \left| \frac{\partial f}{\partial x_i \partial x_k}(j, x) \right|.$$

The operator (2.80) generates a quite specific Markov process on $\{1, \cdots, d\} \times \Sigma_d$ (generally speaking, no longer a chain, since it has a continuous state space). Its second coordinate x evolves according to the deterministic kinetic equations $\dot{x} = Q^T(t, x)x$, independently of the random first coordinate, which, given j, x at time s, evolves according to the time-inhomogeneous Markov chain $J_{s,j}^x(t) \in \{1, \cdots, d\}$ with the Q-matrix

$$Q_{ij}^{dev}(t) = Q_{ij}(t, X_{s,x}(t), u_i^{dev}(t)).$$

Therefore, the transition operators $U^{s,t}$ of this process can be written as

$$U^{s,t} f(j, x) = \mathbf{E} f(J_{s,j}^x(t), X_{s,x}(t)). \tag{2.82}$$

For a function f that does not explicitly depend on x, this simplifies to

$$U^{s,t} f(j, x) = \mathbf{E} f(J_{s,j}^x(t)). \tag{2.83}$$

Theorem 2.7.1. *In the setting of Theorem 2.5.1, let us assume that one distinguished agent in the group of N players deviates from the general rules, moving according to the transition Q-matrix $Q^{dev}(t, x)$, $t \in [0, T]$, satisfying the same regularity assumptions as Q. Let $f(j, x) = f(j)$ not explicitly depend on x. Then*

$$\|(U^{s,t} - U_N^{s,t})f\|_{sup} \leq \frac{(t-s)^{3/2}}{N^{1/2}} C(d, T, \|Q\|_{bLip}, \|Q^{dev}\|_{bLip})\|f\|_{sup},$$

$$0 \leq s \leq t \leq T, \quad (2.84)$$

with a constant C depending on $d, T, \|Q\|_{bLip}, \|Q^{dev}\|_{bLip}$. For smooth Q and Q^{dev},

$$\|(U^{s,t} - U_N^{s,t})f\|_{sup} \leq \frac{(t-s)^2}{N} C(T, \|Q\|_{C^2}, \|Q^{dev}\|_{C^2})\|f\|_{sup}, \tag{2.85}$$

with a constant C depending on $T, \|Q\|_{C^2}, \|Q^{dev}\|_{C^2}$.

Proof. Let us start with the case of smooth Q and Q^{dev}. Using (2.81) and the comparison of propagators formula (2.52), we derive that

$$\|(U^{s,t} - U_N^{s,t})f\|_{sup} = \sup_{j,x} |(U^{s,t} - U_N^{s,t})f(j, x)|$$

$$\leq (t - s) \sup_{r \in [s,t]} \|(L_t^{N,dev} - \Lambda_t^{dev}) U^{r,t} f\|_{sup}$$

$$\leq \frac{t - s}{N} \|Q\| \left(\sup_{r \in [s,t]} \|(U^{r,t} f)^{(2)}\| + 2\|f\| \right). \tag{2.86}$$

Thus we need to estimate

$$\|(U^{r,t} f)^{(2)}\| = \sup_{k,l,j,x} \left| \frac{\partial^2 U^{r,t} f(j, x)}{\partial x_k \partial x_l} \right|,$$

with $U^{s,t}$ given by (2.83) (and with $f(j, x) = f(j)$ not depending on x).

To deal with $U^{s,t}$, it is convenient to fix $s < T$ and x and to consider the auxiliary propagator $U_{[s,x]}^{r,t}$, $s \leq r \leq t \leq T$, of the Markov chain $Y_{r,j}(t)$ in $\{1, \cdots, d\}$ with the Q-matrix $Q^{dev}(t, X_{s,x}(t))$, so that

$$U_{[s,x]}^{r,t} f(j) = \mathbf{E} f(Y_{r,j}(t)), \tag{2.87}$$

and

$$U^{s,t} f(j) = U_{[s,x]}^{s,t} f(j). \tag{2.88}$$

Unlike $U^{s,t}$ acting on functions on $\{1, \cdots, d\} \times \Sigma_d$, the propagator $U_{[s,x]}^{r,t}$ is a propagator of a usual Markov chain, and hence its action satisfies the ODE

$$\frac{d}{dr} U_{[s,x]}^{r,t} f(j) = [Q^{dev}(t, X_{s,x}(t)) U_{[s,x]}^{r,t} f](j) = \sum_k Q_{jk}^{dev}(t, X_{s,x}(t)) (U_{[s,x]}^{r,t} f)(k).$$

To find the derivatives with respect to x, we can use the standard ODE sensitivity results, Proposition 2.2.4 (used in backward time and with the initial condition not depending explicitly on the parameter), yielding

$$\sup_{i,k,x} \left| \frac{\partial}{\partial x_i} U_{[s,x]}^{r,t} f(k) \right| \leq (t - r) \sup_k |f(k)| \sup_{x,i} \left\| \frac{\partial Q^{dev}(t, X_{s,x}(t))}{\partial x_i} \right\| \exp\{2(t - r)\|Q^{dev}\|\}$$

and

$$\sup_{i,j,k} \left| \frac{\partial^2}{\partial x_i \partial x_j} U_{[s,x]}^{r,t} f(k) \right| \leq \exp\{3(t - r)\|Q^{dev}\|\} \sup_k |f(k)|$$

$$\times \left((t - r) \sup_{i,j} \left\| \frac{\partial^2 Q^{dev}(t, X_{s,x}(t))}{\partial x_i \partial x_j} \right\| + (t - r)^2 \sup_i \left\| \frac{\partial Q^{dev}(t, X_{s,x}(t))}{\partial x_i} \right\|^2 \right).$$

Since

$$\frac{\partial Q^{dev}(t, X_{s,x}(t))}{\partial x_i} = \frac{\partial Q^{dev}(t, y)}{\partial y} |_{y = X_{s,x}(t)} \frac{\partial X_{s,x}(t)}{\partial x_i}$$

and similarly for the second derivative, we can use Proposition 2.2.1 to estimate the derivatives of $X_{s,x}(t)$ and thus to obtain $\|(U^{r,t}f)^{(2)}\| \leq C$ with C depending on T, $\|Q\|_{C^2}$, $\|Q^{dev}\|_{C^2}$ and hence (2.85) follows by (2.86).

When Q and Q^{dev} are only Lipschitz, we use the same approximations $\Phi_\delta[Q]$ and $\Phi_\delta[Q^{dev}]$ as in Section 2.5. And as in Section 2.5, we get the rate of convergence of order $td\delta + t^2/(\delta N)$, yielding (2.84) by choosing $\delta = \sqrt{t/N}$. □

Let us extend the result to functions $f(j, x)$ depending on x explicitly.

Theorem 2.7.2. *In the setting of Theorem 2.5.1, let us assume that one distinguished agent in the group of N players deviates from the general rules, moving according to the transition Q-matrix $Q^{dev}(t, x)$, $t \in [0, T]$, satisfying the same regularity assumptions as Q. Then*

$$\|(U^{s,t} - U^{s,t}_N)f\|_{sup} \leq \frac{(t-s)^{1/2}}{N^{1/2}} C(d, T, \|Q\|_{bLip}, \|Q^{dev}\|_{bLip})\|f\|_{bLip},$$

$$0 \leq s \leq t \leq T, \quad (2.89)$$

with a constant C depending on d, T, $\|Q\|_{bLip}$, $\|Q^{dev}\|_{bLip}$. For smooth Q and Q^{dev} and f,

$$\|(U^{s,t} - U^{s,t}_N)f\|_{sup} \leq \frac{(t-s)}{N} C(T, \|Q\|_{C^2}, \|Q^{dev}\|_{C^2})\|f\|_{C^2}, \quad (2.90)$$

with a constant C depending on T, $\|Q\|_{C^2}$, $\|Q^{dev}\|_{C^2}$.

Proof. The only difference between this case and the previous one is the necessity to use Proposition 2.2.4 in full, that is, with the initial condition also depending on the parameter and, in case f is not smooth, to approximate it in the usual way by smooth functions. □

2.8 Stationary Points of Limiting Dynamics and Nash Equilibria

Theorem 2.6.1 suggests that eventually, the controlled Markov evolution will settle down near some stable equilibrium points of the dynamical system (2.71) or (2.72).

Let us deal now specifically with system (2.72). For a subset $I \subset \{1, \cdots, d\}$, let

$$\Omega_I = \{x \in \Sigma_d : x_k = 0, k \in I, \text{ and } R_j(x, b^*(x)) = R_i(x, b^*(x)) \text{ for } i, j \notin I\}.$$

Theorem 2.8.1. *A vector x with nonnegative coordinates is a stationary point of (2.72), that is, it satisfies the system of equations*

$$\sum_i \varkappa x_i x_j [R_j(x, b^*(x)) - R_i(x, b^*(x))] = 0, \quad j = 1, ..., d, \quad (2.91)$$

if and only if $x \in \Omega_I$ *for some* $I \subset \{1, \cdots, d\}$.

Proof. For every I such that $x_k = 0$ for $k \in I$, system (2.91) reduces to the same system but with coordinates $k \notin I$. Hence it is sufficient to prove the result for empty I. In this situation, system (2.91) reduces to

$$\sum_i x_i[R_j(x, b^*(x)) - R_i(x, b^*(x))] = 0, \quad j = 1, ..., d. \tag{2.92}$$

Subtracting the jth and kth equations of this system yields

$$(x_1 + \cdots + x_d)[R_j(x, b^*(x)) - R_k(x, b^*(x))] = 0,$$

and thus

$$R_j(x, b^*(x)) = R_k(x, b^*(x)),$$

as required.   □

So far, we have deduced the dynamics arising from a certain Markov model of interaction. As is known, the internal (not lying on the boundary of the simplex) singular points of the standard replicator dynamics of evolutionary game theory correspond to the mixed-strategy Nash equilibria of the initial game with a fixed number of players (in most examples, just a two-player game). Therefore, it is natural to ask whether a similar interpretation can be given to the fixed points of Theorem 2.8.1. Because of the additional nonlinear mean-field dependence of R on x, the interpretation of x as mixed strategies is not at all clear. However, consider explicitly the following game Γ_N of $N + 1$ players (which was tacitly borne in mind when we discussed dynamics). When the major player chooses the strategy b and each of N small players chooses the state i, the major player receives the payoff $B(x, b, N)$, and each player in state i receives $R_i(x, b)$, $i = 1, \cdots, d$ (as above, with $x = n/N$ and $n = (n_1, \cdots, n_d)$ the realized occupation numbers of all the states). Thus a strategy profile of small players in this game can be specified either by a sequence of N numbers (expressing the choice of the state by each agent) or, more succinctly, by the resulting collection of frequencies $x = n/N$.

As usual (see any text on game theory, e.g., [175] or [194]), one defines a Nash equilibrium in Γ_N as a profile of strategies (x_N, b_N^*) such that for each player, changing its choice of strategy unilaterally would not be beneficial, that is,

$$b_N^* = b^*(x_N, N) = argmax \, B(x_N, b, N),$$

and for all $i, j \in \{1, \cdots, d\}$,

$$R_j(x - e_i/N + e_j/N, b_N^*) \le R_i(x, b_N^*). \tag{2.93}$$

A profile is an ϵ-*Nash equilibrium* if these inequalities hold up to an additive correction term not exceeding ϵ. It turns out that the singular points of (2.72) describe all approximate Nash equilibria for Γ_N in the following precise sense:

Theorem 2.8.2. *Let $R(x, b)$ be Lipschitz continuous in x uniformly in b. Let*

$$\hat{R} = \sup_{i,b} \| R_i(., b) \|_{Lip},$$

and for $I \subset \{1, \cdots, d\}$, let

$$\hat{\Omega}_I = \{ x \in \Omega_I : R_k(x, b^*(x)) \leq R_i(x, b^*(x)) \text{ for } k \in I, i \notin I \}.$$

Then the following assertions hold.

    (i) The limit points of any sequence x_N such that $(x_N, b^*(x_N, N))$ is a Nash equilibrium for Γ_N belong to $\hat{\Omega}_I$ for some I. In particular, if all x_N are internal points of Σ_d, then every limit point belongs to $\Omega_\emptyset$.

    (ii) For all I and $x \in \hat{\Omega}_I$ there exists a $2\hat{R}d/N$-Nash equilibrium $(x_N, b^*(x_N, N))$ to Γ_N such that the difference between any coordinates of x_N and x does not exceed $1/N$ in magnitude.

Proof. (i) Let us consider a sequence of Nash equilibria $(x_N, b^*(x_N, N))$ such that the coordinates of all x_N in I vanish. By (2.93) and the definition of $\hat{R}$,

$$|R_j(x_N, b^*(x_N, N)) - R_i(x_N, b^*(x_N, N))| \leq \frac{2}{N} \hat{R} \qquad (2.94)$$

for all $i, j \notin I$ and

$$R_k(x_N, b^*(x_N, N)) \leq R_i(x_N, b^*(x_N, N)) + \frac{2}{N} \hat{R}, \quad k \in I, i \notin I. \qquad (2.95)$$

Hence $x \in \hat{\Omega}_I$ for every limiting point (x, b).

    (ii) If $x \in \hat{\Omega}_I$, one can construct its $1/N$-rational approximation, namely a sequence $x_N \in \Sigma_d \cap \mathbf{Z}_+^d/N$ such that the difference between any coordinates of x_N and x does not exceed $1/N$ in magnitude. For every such x_N, the profile $(x_N, b^*(x_N, N))$ is a $2\hat{R}d/N$-Nash equilibrium for Γ_N. $\qquad \square$

Theorem 2.8.2 provides a game-theoretic interpretation of the fixed points of the dynamics (2.72), which is independent of any myopic hypothesis used to justify this dynamics. To better illustrate this independence, we can easily extend this theorem to situations in which the solutions to the kinetic equations are not well defined, namely to the case of only continuous $R_j(x)$ (neither Lipschitz not even Hölder). The result can best be expressed in terms of the modulus of continuity of the function R_j, which we define in the following way: $w_j(h; b) = \sup\{|R_j(x, b) - R_j(y, b)|\}$, where sup is over the pairs of x, y that differ only in one coordinate and by an amount not exceeding h. A straightforward extension of the proof of Theorem 2.8.2 yields the following result.

Theorem 2.8.3. *Let $R(x, b)$ be continuous in x uniformly in b, so that $\hat{R}(h) \to 0$ as $h \to 0$, where*

$$\hat{R}(h) = \sup_{i,b} w_i(h; b).$$

Then (i) the limit points of every sequence x_N such that $(x_N, b_N^*(x_N, N))$ is a Nash equilibrium for Γ_N belong to $\hat{\Omega}_I$ for some I; and (ii) for all I and $x \in \hat{\Omega}_I$, there exists a $2d\hat{R}(1/N)$-Nash equilibrium $(x_N, b_N^*(x_N, N))$ to Γ_N such that the difference between any coordinates of x_N and x does not exceed $1/N$ in magnitude.

Theorem 2.8.2 extends also automatically to the case of several major players and dynamics (2.78). For example, let us discuss the case of Lipschitz continuous payoffs. Namely, let us consider the game $\Gamma_{N,K}$ of $N + K$ players, where the major players choose the strategies $b_1, \cdots, b_K$ and each of N small players chooses the state i. The payoffs of the major players are $B(x, b, N)$, and each player in the state i receives $R_i(x, b)$, $i = 1, \cdots, d$. Assume the existence of a continuous branch of Nash equilibria $(b_1^*(x, N), \cdots, b_K^*(x, N))$ having limit (2.77).

Theorem 2.8.4. *Let $R(x, b_1, \cdots, b_K)$ be Lipschitz continuous in x uniformly in $b_1, \cdots, b_K$ and*

$$\hat{R} = \sup_{i, b_1, \cdots, b_K} \|R_i(., b_1, \cdots, b_K)\|_{Lip}.$$

For $I \subset \{1, \cdots, d\}$, let us define Ω_I and $\hat{\Omega}_I$ as above but with $(b_1^*(x), \cdots, b_K^*(x))$ instead of just $b^*(x)$. Then the following statements hold:

(i) The limit points of every sequence x_N such that $(x_N, b_1^*(x_N, N), \cdots, b_K^*(x_N, N))$ is a Nash equilibrium for $\Gamma_{N,K}$ belong to $\hat{\Omega}_I$ for some I.

(ii) For all I and $x \in \hat{\Omega}_I$, there exists a $2\hat{R}d/N$-Nash equilibrium $(x_N, b_1^*(x, N), \cdots, b_K^*(x, N))$ for $\Gamma_{N,K}$ such that the difference between any coordinates of x_N and x does not exceed $1/N$ in magnitude.

Of course, the set of "almost equilibria" Ω may be empty or contain many points. Thus one can naturally pose here the analogue of the question, which is well discussed in the literature on the standard evolutionary dynamics (see [46] and references therein), as to which equilibria can be chosen in the long run (the analogues of stochastically stable equilibria in the sense of [87]) if small mutations are included in the evolution of the Markov approximation.

A distinguished class of stationary points of dynamics (2.72) (and the related Nash equilibria) represent *stable stationary points* (or stable sets of stationary points), characterized by the property that all points starting motion in a neighborhood of such a point (or set of points) remain in this neighborhood forever. The analysis of stable stationary points in the key class of examples will be preformed in Section 2.11. It will be shown there that such stability implies certain "long-term stability," usually for times t of order N, for the approximating dynamics of N players; see Theorem 2.11.2. This theorem is formulated for a class of examples, but it is very general in its nature. In many cases one gets much better results showing that the equilibrium probabilities for the chains of N players have supports in the $1/N$-neighborhood of the set of stable stationary points of the dynamics (2.72).

2.9   The Main Class of Examples: Inspection, Corruption, Security

All models of inspection, corruption, counterterrorist measures, and cybersecurity of Chapter 1 fall in the general class of pressure and resistance games, in which payoffs to small players have the following structure:

$$R_j(x, b) = w_j - p(b) f_j, \tag{2.96}$$

where $w_j > 0$ are profits (or winnings) resulting in applying the jth strategy, and $f_j > 0$ are fines that have to be paid with some probabilities $p(b)$ depending on, and increasing with, the efforts of the principal measured by the budget parameter b. By ordering the strategies of small players, one can assume that

$$w_1 < w_2 < \cdots < w_d, \quad f_1 < f_2 < \cdots < f_d, \tag{2.97}$$

the latter inequalities expressing the natural assumption that fines (risks) increase as one attempts to obtain higher profits. With the principal having essentially opposite interests to the interest of the pool of small players, the payoff of the principal can often be expressed as the weighted average loss of small players with the budget used subtracted:

$$B(x, b) = -b + \varkappa \sum_j x_j (p(b) f_j - w_j) = -b + \varkappa (p(b) \bar{f} - \bar{w}), \tag{2.98}$$

where $\varkappa$ is a constant, and bars denote the averaging with respect to the distribution x.

Remark 12. *Of course, one can think of more general situations with $R_j = w_j - p_j(b) f_j$, with probabilities depending on j or with $R_j = w_j - p(b_j) f_j$ with different budgets used for dealing with different strategies.*

Since p is increasing, $p'(b) > 0$. Moreover, interpreting p as the probability of finding a certain hidden behavior of small players, it is natural to assume that the search can never be perfect, that is, $p(b) \neq 1$, and thus $p : [0, \infty) \to [0, 1)$. Assuming that p is rapidly increasing for small b and that this growth decreases with an increase in b (the analogous assumption to the law of diminishing returns in economics) leads to the conditions that $p'(0) = \infty$ and $p(b)$ is a concave function. Summarizing, the properties of $p : (0, \infty) \to (0, 1)$ that can be naturally assumed for a rough qualitative analysis are as follows: p is a smooth monotone bijection such that

$$p''(b) < 0 \text{ for all } b, \quad p'(0) = +\infty, \quad p'(\infty) = 0. \tag{2.99}$$

Under these assumptions, the function $B(x, b)$ from (2.98) is concave as a function of b, its derivative monotonically decreases from ∞ to -1, and hence there exists a unique maximum $b^* = b^*(x)$ such that $-1 + \varkappa \bar{f} p'(b^*) = 0$, so that

$$b^*(x) = (p')^{-1}(1/\varkappa \bar{f}(x)), \tag{2.100}$$

where $(p')^{-1} : (0, \infty) \to (0, \infty)$ is the (monotonically decreasing) inverse function to p'. Moreover, $b^*(x)$ depends on x only via the average $\bar{f} = \sum_j x_j f_j$, so that

$$b^*(x) = \hat{b}(\bar{f}(x)), \quad \hat{b}(\bar{f}) = (p')^{-1}(1/\varkappa \bar{f}(x)). \tag{2.101}$$

Clearly, the range of possible $b^*(x) = \hat{b}(\bar{f})$ is the interval

$$b^* \in [(p')^{-1}(1/\varkappa f_1), (p')^{-1}(1/\varkappa f_d)]. \tag{2.102}$$

Thus the controlled dynamics (2.72) takes the form

$$\dot{x}_j = x_j[w_j - p(b^*(x)) f_j - (\bar{w} - p(b^*(x)) \bar{f})], \quad j = 1, ..., d. \tag{2.103}$$

The stationary points of this dynamics can be explicitly calculated and their stability analyzed for general classes of dependence of the fines f_j on the profits w_j. The simplest possibility is the *proportional fines* $f_j = \lambda w_j$ with a constant $\lambda > 0$.

Remark 13. *Such fines appear in Russian legislation for punishment arising from tax evasion, that is, in the context of inspection games.*

The next result classifying the stationary points for proportional fines is straightforward.

Proposition 2.9.1. *For the case of proportional fines, the dynamics (2.103) turns into the dynamics*

$$\dot{x}_j = x_j(w_j - \bar{w})(1 - p(b^*(x))\lambda), \quad j = 1, ..., d. \tag{2.104}$$

The stationary points of this dynamics are the vertices of the simplex Σ_d, and if

$$\frac{1}{\lambda} \in p[(p')^{-1}(1/\varkappa f_1), (p')^{-1}(1/\varkappa f_d)], \tag{2.105}$$

then all points of the hyperplane defined uniquely by the equation $\bar{f}(x) = f^*$ with

$$p(\hat{b}(f^*)) = 1/\lambda \qquad (2.106)$$

also represent stationary points.

2.10 Optimal Allocation

So far, our small players have been indistinguishable. However, in many cases the small players can belong to different types. These can be inspectees with various income brackets, levels of danger, or overflow of particular traffic path, or classes of computers susceptible to infection. In this situation the problem for the principal becomes a policy problem, that is, how to allocate efficiently its limited resources. Our theory extends to a setting with various types more or less straightforwardly. We shall touch upon it briefly.

The models of investment policies of Section 1.12 also belong to this class of problems.

Let our players, apart from being distinguished by states $i \in \{1, \cdots, d\}$, also be classified by their types or classes $\alpha \in \{1, \cdots, \mathcal{A}\}$. The state space of the group becomes $\mathbf{Z}_+^d \times \mathbf{Z}_+^{\mathcal{A}}$, the set of matrices $n = (n_{i\alpha})$, where $n_{i\alpha}$ is the number of players of type α in the state i (for simplicity of notation, we identify the state spaces of each type, which is not at all necessary). One can imagine several scenarios of communication between classes, two extreme cases being as follows:

(C1) No communication: the players of different classes can neither communicate nor observe the distribution of states in other classes, so that the interaction between types arises exclusively through the principal.

(C2) Full communication: the players can change both their types and states via pairwise exchange of information, and can observe the total distribution of types and states.

There are many intermediate cases, such as those in which types form a graph (or a network) with edges specifying the possible channels of information. Let us deal here only with cases (C1) and (C2). Starting with (C1), let N_α denote the number of players in class α, and n_α the vector $\{n_{i\alpha}\}$, $i = 1, \cdots, d$. Let $x_\alpha = n_\alpha/N_\alpha$,

$$x = (x_{i\alpha}) = (n_{i\alpha}/N_\alpha) \in (\Sigma_d)^{\mathcal{A}},$$

and let $b = (b_1, \cdots, b_{\mathcal{A}})$ be the vector of the allocation of resources of the principal, which may depend on x. Assuming that the principal uses the optimal policy

$$b^*(x) = argmax\, B(x, b) \qquad (2.107)$$

arising from some concave (in the second variable) payoff function B on $(\Sigma_d)^{\mathcal{A}} \times \mathbf{R}^{\mathcal{A}}$, the generator of the controlled Markov process becomes

$$L_{b*,N} f(x) = \sum_{\alpha=1}^{\mathcal{A}} N_\alpha \varkappa_\alpha \sum_{i,j} x_{i\alpha} x_{j\alpha}$$

$$\times [R_j^\alpha(x_\alpha, b^*(x)) - R_i^\alpha(x_\alpha, b^*(x))]^+ [f(x - e_i^\alpha/N_\alpha + e_j^\alpha/N_\alpha) - f(x)], \quad (2.108)$$

where e_i^α is now the standard basis in $\mathbf{R}^d \times \mathbf{R}^{\mathcal{A}}$. Passing to the limit as $N \to \infty$ under the assumption that

$$\lim_{N \to \infty} N_\alpha/N = \omega_\alpha$$

with some constants ω_α, we obtain a generalization of (2.72) in the form

$$\dot{x}_{j\alpha} = \varkappa_\alpha \omega_\alpha \sum_i x_{i\alpha} x_{j\alpha} [R_j^\alpha(x_\alpha, b^*(x)) - R_i^\alpha(x_\alpha, b^*(x))], \qquad (2.109)$$

for $j = 1, ..., d$ and $\alpha = 1, \cdots, \mathcal{A}$, coupled with (2.107).

In case (C2), $x = (x_{i\alpha}) \in \Sigma_{d\alpha}$, the generator becomes

$$L_{b*,N} f(x) = \sum_{\alpha,\beta=1}^{\mathcal{A}} N\varkappa \sum_{i,j} x_{i\beta} x_{j\alpha}$$

$$\times [R_j^\alpha(x, b^*(x)) - R_i^\beta(x, b^*(x))]^+ [f(x - e_i^\beta/N_\alpha + e_j^\alpha/N_\alpha) - f(x)], \quad (2.110)$$

and the limiting system of differential equations

$$\dot{x}_{j\alpha} = \varkappa \sum_{i,\beta} x_{i\beta} x_{j\alpha} [R_j^\alpha(x, b^*(x)) - R_i^\beta(x, b^*(x))]. \qquad (2.111)$$

2.11 Stability of Stationary Points and Its Consequences

As noted above, an important class of stationary points represents stable points.

Let us show that the hyperplane of fixed points (2.106) is *stable* under the dynamics (2.104), that is, if the dynamics starts in a sufficiently small neighborhood of this hyperplane, then it would remain there forever. We shall do the by the method of Lyapunov functions, showing that the function

$$V(x) = (1 - \lambda p(b^*(x)))^2$$

is a *Lyapunov function* for the dynamics (2.104), meaning that $V(x) = 0$ only in the plane (2.106) and $\dot{V}(x) \le 0$ everywhere whenever x evolves according to (2.104).

Proposition 2.11.1. *If x evolves according to (2.104), then $\dot{V}(x) < 0$ for all x that are not vertices of Σ_d and do not belong to the plane (2.106), implying that the neighborhoods $\{x : V(x) < v\}$ are invariant under (2.104) for all v.*

Proof. We have

$$\frac{d}{dt}V(x) = \frac{\partial V}{\partial x}\dot{x} = -2\lambda(1 - \lambda p(b^*(x)))^2 p'(\hat{b}(\bar{f}))\hat{b}'(\bar{f})\sum_j x_j f_j(w_j - \bar{w})$$

$$= -2\lambda^2(1 - \lambda p(b^*(x)))^2 p'(\hat{b}(\bar{f}))\hat{b}'(\bar{f})(\sum_j x_j w_j^2 - \bar{w}^2)$$

$$= -2\lambda^2 V(x) p'(\hat{b}(\bar{f}))\hat{b}'(\bar{f})(\sum_j x_j w_j^2 - \bar{w}^2) < 0,$$

since $\sum_j x_j w_j^2 - \bar{w}^2$ is the variance of the random variable w taking values w_j with probabilities x_j, which is always nonnegative. $\qquad\square$

The statement of this proposition can be essentially improved to yield a full portrait of the dynamics (2.104).

Theorem 2.11.1. *(i) If the plane of stationary points (2.106) exists, that is, condition (2.105) holds, then this plane is the global attractor for the dynamics (2.104) outside vertices of Σ_d: for every x that is not a vertex of Σ_d, $X_x(t)$ approaches this plane as $t \to \infty$. Moreover, the simplex Σ_d can be decomposed into two invariant sets with $1 - p(b^*(x))\lambda > 0$ and $1 - p(b^*(x))\lambda < 0$.*
(ii) If

$$1/\lambda \geq p[(p')^{-1}(1/\varkappa f_d)],$$

then the pure strategy of maximal activity $j = d$ is a global attractor: for every x that is not a vertex of Σ_d, $X_x(t)$ approaches the point $(0, \cdots, 0, 1)$ as $t \to \infty$.
(iii) If

$$1/\lambda \leq p[(p')^{-1}(1/\varkappa f_1)],$$

then the pure strategy of minimal activity $j = 1$ is a global attractor: for every x that is not a vertex of Σ_d, $X_x(t)$ approaches the point $(1, 0, \cdots, 0)$ as $t \to \infty$.

Proof. (i) Since for every x that is not a vertex of Σ_d, $X_x(t)$ can also never become a vertex, it follows that $(d/dt)V(X_x(t)) < 0$ on the whole trajectory. Hence there exists a limit of $V(X_x(t))$ as $t \to \infty$. A straightforward argument by contradiction shows that this limit must be zero.

(ii) In this case, $1 - p(b^*(x))\lambda > 0$ in the whole of Σ_d, and the minimum of V in Σ_d occurs at $(0, \cdots, 0, 1)$.

(iii) In this case, $1 - p(b^*(x))\lambda < 0$ in the whole of Σ_d, and the minimum of V in Σ_d occurs at $(1, 0, \cdots, 0)$. $\qquad\square$

Let us describe a consequence of the stability of the hyperplane (2.106) for the behavior of the dynamics U_N^t of a finite number of players.

Theorem 2.11.2. *If $V(x) \leq v$, then with high probability, the points $X_x^N(t)$ remain in the neighborhood $\{V(y) \leq rv\}$ for large r and times t much less than N. Namely,*

$$\mathbf{P}(V(X_x^N(t)) > rv) \leq \frac{1}{r}\left(1 + \frac{\epsilon}{v}\|V^{(2)}\|\,\|Q\|\right) \tag{2.112}$$

for $t \leq \epsilon N$.

Proof. By (2.39),

$$\|L_t^N V - \Lambda_t V\| \leq \frac{1}{N}\|V^{(2)}\|\,\|Q\|.$$

Hence by Proposition 2.11.1,

$$L_t^N V(x) \leq \frac{1}{N}\|V^{(2)}\|\,\|Q\|$$

for all x. Consequently,

$$\frac{d}{dt}U_N^t V(x) = L_N^t V(x) \leq \frac{1}{N}\|V^{(2)}\|\,\|Q\|$$

for all x, and therefore

$$U_N^t V(x) \leq V(x) + \frac{t}{N}\|V^{(2)}\|\,\|Q\|.$$

Consequently, if $t \leq \epsilon N$ for some small ϵ and $V(x) \leq v$, then

$$\mathbf{E}V(X_x^N(t)) \leq v + \epsilon\|V^{(2)}\|\,\|Q\|,$$

implying (2.112) by Markov's inequality.   □

This theorem is very rough and has an almost straightforward extension to the general stable stationary points of the dynamics (2.71). For the particular case of tye dynamics (2.104), the full description of the large-time behavior of the Markov chains $X_x^N(t)$ is available.

Theorem 2.11.3. *(i) Under the condition of Theorem 2.11.1 (ii) or (iii), the points $(0, \cdots, 0, 1)$ and $(1, 0, \cdots, 0)$, respectively, are the absorbing points for Markov chains $X_x^N(t)$ for any N: starting from any profile of strategies, the trajectory almost surely reaches this point in finite time and remains there forever. (ii) Under the condition of Theorem 2.11.1 (i), the $2/N$-neighborhood of the plane (2.106) is absorbing for the Markov chains $X_x^N(t)$ for any N: starting from any profile of strategies, the trajectory almost surely enters this neighborhood in a finite time and remains there* *forever. It follows, in particular, that these Markov chains have stationary distributions supported on the states belonging to the 2/N-neighborhoods of the plane (2.106).*

Proof. The generator (2.31) of the Markov chain $X_x^N(t)$ for transitions (1.10), (2.96) is

$$L_t^N f(x) = N \sum_{i=1}^{d} \sum_{j=1}^{d} x_i x_j [w_j - w_i - p(b^*(x))(f_j - f_i)]^+$$

$$[f(x - e_i/N + e_j/N) - f(x)], \quad x \in \mathbf{Z}_+^d/N, \tag{2.113}$$

which, for the case of proportional fines, takes the form

$$L_t^N f(x) = N \sum_{i=1}^{d} \sum_{j=1}^{d} x_i x_j [(w_j - w_i)(1 - p(b^*(x))\lambda)]^+$$

$$[f(x - e_i/N + e_j/N) - f(x)]. \tag{2.114}$$

Therefore, when $1 - p(b^*(x))\lambda < 0$, only transitions decreasing activity, $i \to j < i$, are allowed. Conversely, when $1 - p(b^*(x))\lambda > 0$, only transitions increasing activity, $i \to j > i$ are allowed. This implies statement (i), since in this case, the sign of $1 - p(b^*(x))\lambda$ is constant on the whole of Σ_d.

In case (ii), the transitions from points on the "upper set" and "lower set,"

$$\Sigma_d^+ = \{x : 1 - p(b^*(x))\lambda < 0\}, \quad \Sigma_d^- = \{x : 1 - p(b^*(x))\lambda > 0\},$$

decrease or increase the activity, respectively. As long as these points are outside a 2/N-neighborhood of the plane (2.106), any single transition cannot jump from the upper to the lower set or conversely, but it decreases the distance of the states to the plane (2.106). Thus the points continue to move until they enter this neighborhood. □

The analysis of system (2.103) for nonproportional fines is given in [125]. In cases of convex and concave dependence of fines on the profits, all fixed points turn out to be isolated and supported by only two pure strategies.

These results lead to clear practical conclusions showing how manipulation of the structure of fines (often imposed by the principal) can yield the desired distribution of the stationary points of the related dynamics and hence the equilibria of the corresponding games. The choice of the fine structure can be looked at as a kind of *mechanism design* of the principal. For instance, in the case of proportional fines considered above, an appropriate choice of the coefficient of proportionality can lead the whole pool of small players to act on the maximal or on the minimal level of activity.

2.12   Stability for Two-State Models

Of course, one expects the simplest evolutions to occur for two-state models. In fact, under reasonable assumptions, the behavior of both limiting and approximating dynamics can be fully sorted out in this case.

Assume that the state space of small players consists of only two strategies. Then the rates are specified just by two numbers Q_{12} and Q_{21}, describing transitions from the first state to the second and conversely, and the overall distribution is specified by one number, $x = n/N$, the fraction of players using the first strategy. In dealing with time-independent transitions, the limiting dynamics (2.40) reduces to the following single equation:

$$\dot{x} = (1 - x)Q_{21}(x) - xQ_{12}(x). \tag{2.115}$$

Assume now that the interval $[0, 1]$ is decomposed into a finite number of regions with the directions of preferred transitions alternating between them. Namely, let

$$0 = a_0 < a_1 < \cdots < a_{k-1} < a_k = 1, \quad I_k = (a_{k-1}, a_k),$$

and let $Q_{21} > 0$, $Q_{12} = 0$ in I_k with even k, and $Q_{21} = 0$, $Q_{12} > 0$ in I_k with odd k. Assume also that $Q_{12}(x)$ and $Q_{21}(x)$ are Lipschitz continuous and vanish on the boundary points $x = 0$ and $x = 1$.

The minority game of Section 1.10 fits into these assumptions with just two intervals: $I_0 = (0, 1/2)$, $I_1 = (1/2, 1)$.

The fixed points of the dynamics are the points of "changing interests": $x = a_j$, $j = 0, \cdots, k$. It is clear that the points a_k with even (respectively odd) k are stable (respectively unstable) fixed points of the dynamics (2.115). Therefore, if the dynamics starts at $x \in I_k$ with even (respectively odd) k, the solution $X_x(t)$ of (2.115) will tend to the left point a_k (respectively right point a_{k+1}) of I_k as $t \to \infty$.

What is interesting is that in this case, the same behavior can be seen to hold for the approximating systems of N agents evolving according to the corresponding Markov chain $X_x^N(t)$ with the generator

$$L_t^N f(x) = xQ_{12}(x)N[f(x - 1/N) - f(x)] + (1 - x)Q_{21}(x)N[f(x + 1/N) - f(x)]. \tag{2.116}$$

Proposition 2.12.1. *Under the above assumptions and for all sufficiently large N, the Markov chain $X_x^N(t)$ starting at a point $x = n/N \in (a_k, a_{k+1})$ moves to the left (respectively right) if k is even (respectively odd), reaches a $1/N$-neighborhood of a_k (respectively a_{k+1}) in finite time, and remains in that neighborhood forever thereafter.*

Proof. This is straightforward.   □

2.13 Discontinuous Transition Rates

An interesting topic to mention is the study of nonlinear Markov chains with discontinuous rates $Q_{ij}(x)$. Especially important is the case in which the whole simplex Σ_d is decomposed into a finite union of closed domains $\Sigma_d = \cup_{j=1}^{k} \bar{X}_j$ such that the interiors X_j of $\bar{X}_j$ do not intersect and $Q_{ij}(x)$ are Lipschitz continuous in each X_j with the continuous extension to $\bar{X}_j$. Such a system naturally arises in MFG forward–backward systems (see Chapter 6) with a finite set of controls.

We shall not develop here the full theory of discontinuous rates, but we shall establish a basis for it by looking at the specific case of the two-state models of the previous section. Namely, using the notation of that section, let us assume now that Q_{21} and Q_{12} are not continuous through the switching points a_k. As the simplest assumption, let us choose these rates to be constant:

$$Q_{21}(x) = \begin{cases} 1, & x \in I_k, \ k \text{ even,} \\ 0, & x \in I_k, \ k \text{ odd,} \end{cases} \qquad Q_{12}(x) = \begin{cases} 1, & x \in I_k, \ k \text{ odd,} \\ 0, & x \in I_k, \ k \text{ even.} \end{cases} \qquad (2.117)$$

The kinetic equations (2.115) have to be understood now in the sense of Filippov, see [84, 85], which means in this case that instead of the equation, we look at the kinetic differential inclusion

$$\dot{x} \in F(x), \quad F(x) = \begin{cases} (1-x)Q_{21}(x) - xQ_{12}(x), & x \notin \{a_0, \cdots, a_k\} \\ [-a_j, 1-a_j], & x = a_j, \end{cases} \qquad (2.118)$$

so that away from points a_j, the inclusion coincides with the equation (2.115). The solution of this inclusion is not unique only if started from the points a_k with odd k (i.e., from the unstable equilibria). Discarding these initial points, we see that starting from $x_0 = a_k$ with even k, the unique solution to (2.118) remains in a_k. Starting from $x_0 \in I_k$ with even k, the unique solution to (2.118) moves according to $\dot{x} = 1 - x$ until it reaches a_k, where it remains forever, that is,

$$x(t) = \begin{cases} e^{-t}(x_0 - 1) + 1, & t \le t_0 = \ln(1 - x_0) - \ln(1 - a_k), \\ a_k, t \ge t_0. \end{cases}$$

Similarly, starting from $x_0 \in I_k$ with odd k, the unique solution to (2.118) moves according to $\dot{x} = -x$ until it reaches a_{k-1}, where it remains forever.

Comparing this behavior of the kinetic inclusion with the behavior of the approximating Markov chain $X_x^N(t)$ (it was described in Section 2.12 and remains the same for the present discontinuous case), we obtain the following.

Proposition 2.13.1. *Under the above assumptions, the Markov chain $X_x^N(t)$ starting at a point $x = k/N$ that does not equal any a_k with odd k converges weakly to the unique solution of the differential inclusion (2.118) starting from the same point. Moreover, this convergence is even uniform in all times.*

2.14 Nonexponential Waiting Times and Fractional LLN

The key point in the probabilistic description of both Markov and nonlinear Markov chains was the assumption that the waiting time between transitions is an exponential random variable. Here we shall touch upon another extension of the results of Sections 2.4 and 2.5, in which more general waiting times are allowed. The characteristic feature of the exponential waiting times is its memoryless property: the distribution of the remaining time does not depend on the time one has waited so far. Thus refuting exponential assumptions leads necessarily to some effects of memory and to non-Markovian evolutions. Such evolutions are usually described by fractional in time differential equations, which have drawn much attention in the recent scientific literature. Let us obtain the dynamic LLN for interacting multiagent systems for the case of nonexponential waiting times with power tail distributions. As one can expect, this LLN will no longer be deterministic, and our exposition in this section will be more demanding mathematically than in other parts of the book.

Let us say that a positive random variable τ has the probability law $\mathbf{P}$ on $[0, \infty)$ with a *power tail of index* α if

$$\mathbf{P}(\tau > t) \sim \frac{1}{t^\alpha}$$

for large t, that is, the ratio of the l.h.s. and the r.h.s. tends to 1 as $t \to \infty$. For every $\alpha \in (0, 1)$, such a random variable has infinite expectation. Nevertheless, the index α characterizes in some sense the length of the waiting time (and thus the rate at which the sequence of independent identically distributed random variables of this kind evolves), since the waiting time increases with a decrease in α.

As the exponential tails, the power tails are well suited for taking minima. Namely, if $\tau_j, j = 1, \cdots, d$, are independent variables with a power tail of indices α_i, then $\tau = \min(\tau_1, \cdots, \tau_d)$ is clearly a variable with a power tail of index $\alpha = \alpha_1 + \cdots + \alpha_d$.

Assume for simplicity that the family $\{Q(x)\}$ does not depend on time. Then in full analogy with the case of exponential times of Sections 1.2 and 2.3, let us assume that the waiting time for the agents of type i to change the strategy has a power tail with index $\alpha_i(x) = \alpha x_i |Q_{ii}|$ with some fixed $\alpha > 0$. Consequently, the minimal waiting time of all agents will have the probability law $P_x(dy)$ with a tail of index

$$\alpha A(x) = \sum \alpha_i(x) = \alpha \sum_i x_i |Q_{ii}(x)|.$$

Our process with power tail waiting times can thus be described probabilistically as follows. Starting from any time and current state n, or $x = n/N$, we wait a random waiting time τ, which has a power tail with index $A(x)$. After this time, the choice of the pair (i, j) such that the transition of one agent from state i to state j occurs is carried out with probability (1.2):

$$\mathbf{P}(i \to j) = \frac{n_i Q_{ij}(n/N)}{\sum_k n_k |Q_{kk}(n/N)|} = \frac{x_i Q_{ij}(x)}{A(x)}. \tag{2.119}$$

After such a jump, the process continues from the new state $x - e_i/N + e_j/N$ in the same way as previously from x.

The easiest way to see what kind of LLN one can expect under this setting and appropriate scaling is to lift the non-Markovian evolution on the space of sequences $x = (x_1, \cdots, x_d) \in \Sigma_d \cap \mathbf{Z}_+^d/N$ to a discrete-time Markov chain by considering the total waiting time s as an additional space variable. Namely, let us consider the Markov chain $(X_{x,s}^{N,\tau}, S_{x,s}^{N,\tau})(k\tau)$ on $(\Sigma_d \cap \mathbf{Z}_+^d/N) \times \mathbf{R}_+$ with the jumps occurring at discrete times $k\tau$, $k \in \mathbf{N}$, such that the process at a state (x, s) at time τk jumps to $(x - e_i/N + e_j/N, s + \tau^{1/\alpha A(x)} y)$ with the probability distribution

$$P_x(dy)\mathbf{P}(i \to j) = P_x(dy)\frac{\tau x_i Q_{ij}(x)}{A(x)}. \tag{2.120}$$

The key to the reasonable scaling is to choose $\tau = 1/N$. We shall denote the corresponding Markov chain by $(X_{x,s}^\tau, S_{x,s}^\tau)(k\tau)$. Its transition operator is

$$U^\tau f(x, s) = \int P_x(dy) \sum_i \sum_{j \neq i} \frac{x_i Q_{ij}(x)}{A(x)} f\left(x - \frac{e_i}{N} + \frac{e_j}{N}, s + \tau^{1/\alpha A(x)} y\right). \tag{2.121}$$

What we are interested in is not exactly this chain, but the value of its first coordinate $X_{x,s}^\tau$ at the time $k\tau$, when the total waiting time $S_{x,s}^\tau(k\tau)$ (which is our real time, unlike the artificial time s) reaches t, that is, at the time

$$k\tau = T_{x,s}^\tau(t) = \inf\{m\tau : S_{x,s}^\tau(m\tau) \geq t\},$$

so that $T_{x,s}^\tau$ is the inverse process to $S_{x,s}^\tau$. Thus the *scaled mean-field interacting system of agents with a power tail waiting time between jumps* is the (non-Markovian) process

$$\tilde{X}_{x,s}^\tau(t) = X_{x,s}^\tau(T_{x,s}^\tau(t)). \tag{2.122}$$

Let us see what happens in the limit $\tau \to 0$, or equivalently $N = 1/\tau \to \infty$. Since the transitions in $\Sigma_d \cap \mathbf{Z}_+^d/N$ do not depend on the waiting times, the first coordinate of the chain $(X_{x,s}^\tau, S_{x,s}^\tau)(k\tau)$ is itself a discrete-time Markov chain, which converges (by Theorems 2.4.3 and 2.5.2) to the deterministic process described by the solutions $X_{x,s}^A(t)$ to the kinetic equations $\dot{x} = xQ(x)/A(x)$ with the initial condition x at time s. Thus we can expect that the limit of the process (2.122), which is the LLN we are looking for, is the process obtained from the characteristics $X_{x,0}^A(t)$ evaluated at some random time t.

Let us see more precisely what happens with the whole chain $(X_{x,s}^\tau, S_{x,s}^\tau)(k\tau)$. It is well known (see, e.g., Theorem 8.1.1 of [137]) that if the operator

$$\Lambda^A f = \lim_{\tau \to 0} \frac{1}{\tau}(U^\tau f - f) \tag{2.123}$$

is well defined and generates a Feller process, then this Feller process represents the weak limit of the scaled chain with the transitions $[U^\tau]^{[t/\tau]}$, where $[t/\tau]$ denotes the integer part of the number t/τ. Thus we need to calculate (2.123). We have

$$\frac{1}{\tau}(U^\tau f - f) = \frac{1}{\tau}\int P_x(dy) \sum_i \sum_{j\neq i} \frac{x_i Q_{ij}(x)}{A(x)}\left[f\left(x - \frac{e_i}{N} + \frac{e_j}{N}, s + \tau^{1/\alpha A(x)}y\right) - f(x,s)\right]$$

$$= \frac{1}{\tau}\int P_x(dy)\left[f(x, s + \tau^{1/\alpha A(x)}y) - f(x,s)\right]$$

$$+ \frac{1}{\tau}\sum_i \sum_{j\neq i} \frac{x_i Q_{ij}(x)}{A(x)}\left[f\left(x - \frac{e_i}{N} + \frac{e_j}{N}, s\right) - f(x,s)\right] + R, \qquad (2.124)$$

where the error term equals

$$R = \frac{1}{\tau}\int P_x(dy) \sum_i \sum_{j\neq i} \frac{x_i Q_{ij}(x)}{A(x)}$$

$$\times \left[\left(f\left(x - \frac{e_i}{N} + \frac{e_j}{N}, s + \tau^{1/\alpha A(x)}y\right) - f\left(x, s + \tau^{1/\alpha A(x)}y\right)\right)\right.$$

$$\left. - \left(f\left(x - \frac{e_i}{N} + \frac{e_j}{N}, s\right) - f(x,s)\right)\right]$$

$$= \frac{1}{\tau}\int P_x(dy) \sum_i \sum_{j\neq i} \frac{x_i Q_{ij}(x)}{A(x)}\left[g(x, s + \tau^{1/\alpha A(x)}y) - g(x,s)\right],$$

with

$$g(x,s) = f\left(x - \frac{e_i}{N} + \frac{e_j}{N}, s\right) - f(x,s).$$

From the calculations of Section 2.4, we know that the second term in (2.124) converges to $\Lambda f(x,s)/A(x)$, with

$$\Lambda f(x,s) = \sum_{i=1}^d \sum_{j\neq i} x_i Q_{ij}(x)\left[\frac{\partial f}{\partial x_j} - \frac{\partial f}{\partial x_i}\right](x,s)$$

$$= \sum_{k=1}^d \sum_{i\neq k}[x_i Q_{ik}(x) - x_k Q_{ki}(x)]\frac{\partial f}{\partial x_k}(x,s),$$

whenever f is continuously differentiable in x.

By Theorem 8.3.1 of [137] (see also [133]), the first term in (2.124) converges to

$$\alpha A(x)\int_0^\infty \frac{f(x, s + y) - f(x, s)}{y^{1+\alpha A(x)}}dy$$

whenever f is continuously differentiable in s.

To estimate the term R, we note that if $f \in C^1(\Sigma_d \times \mathbf{R}_+)$, then $g(x)$ is uniformly bounded by $1/N$ and $|\partial g/\partial s|$ is uniformly bounded. Hence applying again Theorem 8.3.1 of [137] to R, we obtain that $R/\tau \to 0$ as $\tau \to 0$. Hence it follows that for $f \in C^1(\Sigma_d \times \mathbf{R}_+)$,

$$
\Lambda^A f(x, s) = \lim_{\tau \to 0} \frac{1}{\tau} (U^\tau f - f)(x, s)
$$
$$
= \alpha A(x) \int_0^\infty \frac{f(x, s + y) - f(x, s)}{y^{1+\alpha A(x)}} dy + \Lambda f(x, s)/A(x).
$$
(2.125)

Thus we see again that the first coordinate of the chain $(X_{x,s}^\tau, S_{x,s}^\tau)(k\tau)$ converges to the deterministic process described by the solutions $X_{x,s}^A(t)$ (starting in x at time s) to the kinetic equations $\dot{x} = x Q(x)/A(x)$. The second coordinate of $(X_{x,s}^\tau, S_{x,s}^\tau)(k\tau)$ depends on the first coordinate. For $s = 0$ it converges to the similar stable process with the time-dependent family of generators

$$
\Lambda_S^A g(t) = \alpha A(X_{x,0}^A(t)) \int_0^\infty \frac{g(t + y) - g(t)}{y^{1+\alpha A(X_{x,0}^A(t))}} dy.
$$
(2.126)

Deriving the convergence of the processes from the convergence of the generators as in Section 2.4 and using (2.122), we arrive at the following result.

Theorem 2.14.1. *If $Q(x)$ is a Lipschitz continuous function, then the process of the mean-field interacting system of agents $\tilde{X}_{x,s}^\tau(t)$ given by (2.122) converges in distribution to the process*

$$
\tilde{X}_{x,s}(t) = X_{x,s}^A(T_x^t),
$$
(2.127)

where T_x^t is the random time when the stable-like process generated by (2.126) and started at s reaches the time t. Moreover, from the probabilistic interpretation of the generalized Caputo-type derivative of mixed orders $A(x)$ (see [139]), it follows that the evolution of averages $f(x, s) = \mathbf{E} f(\tilde{X}_{x,s}(t))$ satisfies the following generalized fractional differential equation:

$$
D_{t-*}^{A(x)} f(x, s) = \frac{1}{A(x)} \sum_{k=1}^d \sum_{i \neq k} [x_i Q_{ik}(x) - x_k Q_{ki}(x)] \frac{\partial f}{\partial x_k}(x, s), \quad s \in [0, t],
$$
(2.128)

with the terminal condition $f(x, t) = f(x)$, where the left fractional derivative acting on the variable $s \leq t$ of $f(x, s)$ is defined as

$$
D_{t-*}^{A(x)} g(s) = \alpha A(x) \int_0^{t-s} \frac{g(s + y) - g(s)}{y^{1+\alpha A(x)}} dy + \alpha A(x)(g(t) - g(s)) \int_{t-s}^\infty \frac{dy}{y^{1+\alpha A(x)}} dy.
$$
(2.129)

As usual with the nondeterministic LLN, the limiting process depends essentially on the scaling and the choice of approximations. Instead of modeling the total waiting time by a random variable with tail of order $t^{-\alpha A(x)}$, we can equally well model this time by a random variable with tail of order $A^{-1}(x)t^{-\alpha}$, which will yield similar results with the mixed fractional derivative of a simpler form

$$\tilde{D}_{t-*}^{A(x)} g(s) = \frac{\alpha}{A(x)} \int_0^{t-s} \frac{g(s+y) - g(s)}{y^{1+\alpha}} dy + \frac{\alpha(g(t) - g(s))}{A(x)} \int_{t-s}^{\infty} \frac{dy}{y^{1+\alpha}} dy$$

$$= \frac{\alpha}{A(x)} \int_0^{t-s} \frac{g(s+y) - g(s)}{y^{1+\alpha}} dy + \frac{g(t) - g(s)}{A(x)(t-s)^{\alpha}}. \qquad (2.130)$$

However, any scaling of the jump times will lead to the process obtained from the characteristics (solutions of $\dot{x} = x Q(x)$) via a certain random time change.

The situation changes drastically if the family $Q(t, x)$ is time-dependent. Then the partial decoupling (possibility to consider the first coordinate ($X_{x,s}^\tau$ independent of the second one) does not occur, and formula (2.127) does not hold. Nevertheless, the fractional equation (2.128) extends directly to this case, turning into the equation

$$D_{t-*}^{A(x)} f(x, s) = \frac{1}{A(x)} \sum_{k=1}^{d} \sum_{i \neq k} [x_i Q_{ik}(s, x) - x_k Q_{ki}(s, x)] \frac{\partial f}{\partial x_k}(x, s), \quad s \in [0, t].$$

$$(2.131)$$

Unlike (2.127), its solution can be represented probabilistically either via the technique of Dynkin's martingale (see [139]) or via the chronological Feynmann–Kac formula (see [141]).

Chapter 3
Dynamic Control of Major Players

Here we begin to exploit another setting for major player behavior. We shall assume that the major player has some planning horizon with both running and (in case of a finite horizon) terminal costs. For instance, running costs can reflect real spending, while terminal costs can reflect some global objective, such as reducing the overall crime level by a specified amount. This setting will lead us to a class of problems that can be called Markov decision (or control) processes (for the principal) on the evolutionary background (of permanently varying profiles of small players). We shall obtain the corresponding LLN limit for both discrete and continuous time. For discrete time, the LLN limit turns into a deterministic multistep control problem in the case of one major player, and to a deterministic multistep game between major players in the case of several such players. In continuous-time modeling, the LLN limit turns into a deterministic continuous-time dynamic control problem in the case of one major player and to a deterministic differential game in the case of several major players. We analyze the problem with both finite and infinite horizons. The latter case is developed both for the payoffs with discounting and without it. The last version leads naturally to the so-called turnpike behavior of both limiting and prelimiting (finite N) evolutions. The theory of this chapter also has an extension arising from nonexponential waiting times, leading to the control fractional dynamics in the spirit of Section 2.14, but we do not touch on this extension here (see, however, [151]).

3.1 Multistep Decision-Making of the Principal

As above, we shall work with the case of a finite state space of small players, so that the state space of the group is given by vectors $x^{\cdot} = (n_1, \cdots, n_d)/N$ from the lattice $\mathbf{Z}_+^d/N$.

© Springer Nature Switzerland AG 2019

V. N. Kolokoltsov and O. A. Malafeyev, *Many Agent Games in Socio-economic Systems: Corruption, Inspection, Coalition Building, Network Growth, Security*, Springer Series in Operations Research and Financial Engineering, https://doi.org/10.1007/978-3-030-12371-0_3

Starting with the discrete-time case, we denote by $X_N(t, x, b)$ the Markov chain generated by (2.69) with a fixed b taken from a certain convex compact subset of a Euclidean space, that is, by the operator

$$L^{b,N} f(x) = N \sum_i x_i Q_{ij}(x, b) \left[f \left(x - \frac{e_i}{N} + \frac{e_j}{N} \right) - f(x) \right], \qquad (3.1)$$

and starting in $x \in \mathbf{Z}_+^d / N$ at the initial time $t = 0$.

Let us assume that the principal is updating its strategy at discrete times $\{k\tau\}$, $k = 0, 1, \cdots, n - 1$, with some fixed $\tau > 0$, $n \in \mathbf{N}$, aiming at finding a strategy $\pi = \{b_0, b_1, \cdots, b_{n-1}\}$ maximizing the reward

$$V_n^{\pi,N}(x(N)) = \mathbf{E}_{N,x(N)} \left[\tau B(x_0, b_0) + \cdots + \tau B(x_{n-1}, b_{n-1}) + V(x_n) \right], \qquad (3.2)$$

where B and V are given functions (the running and the terminal payoff), $x(N) \in \mathbf{Z}_+^d / N$ also given,

$$x_k = X_N(\tau, x_{k-1}, b_{k-1}), \quad k = 1, 2, \cdots,$$

and $b_k = b_k(x_k)$ are measurable functions of the current state $x = x_k$ ($\mathbf{E}_{N,x(N)}$ denotes the expectation specified by such a process).

By basic dynamic programming (see, e.g., [112] or [144]), the maximal rewards $V_n^N(x(N)) = \sup_\pi V_n^{\pi,N}(x(N))$ at different times k are linked by the optimality equation $V_k^N = S[N]V_{k-1}^N$, where the Shapley operator $S[N]$ (sometimes referred to as the Bellman operator) is defined by the equation

$$S[N]V(x) = \sup_b \left[\tau B(x, b) + \mathbf{E}V(X_N(\tau, x, b)) \right], \qquad (3.3)$$

so that V_n can be obtained by the nth iteration of the Shapley operator:

$$V_n^N = S[N]V_{n-1}^N = S^n[N]V. \qquad (3.4)$$

We are again interested in the law of large numbers limit $N \to \infty$, where we expect the limiting problem for the principal to be the maximization of the reward

$$V_n^\pi(x_0) = \tau B(x_0, b_0) + \cdots + \tau B(x_{n-1}, b_{n-1}) + V_0(x_n), \qquad (3.5)$$

where

$$x_0 = \lim_{N \to \infty} x(N) \qquad (3.6)$$

(which is supposed to exist) and

$$x_k = X(\tau, x_{k-1}, b_{k-1}), \quad k = 1, 2, \cdots, \qquad (3.7)$$

with $X(t, x, b)$ denoting the solution to the characteristic system (or kinetic equations)

$$\dot{x}_k = \sum_{i=1}^{d} x_i Q_{ik}(x, b(x)), \quad k = 1, ..., d, \tag{3.8}$$

with the initial condition x at time $t = 0$, or in the pressure and resistance framework,

$$\dot{x}_j = \sum_i \varkappa x_i x_j [R_j(x, b) - R_i(x, b)], \quad j = 1, ..., d. \tag{3.9}$$

Again by dynamic programming, the maximal reward in this problem, $V_n(x) = \sup_\pi V_n^\pi(x)$, $\pi = \{b_k\}$, is obtained by the iterations of the corresponding Shapley operator, $V_n = S^n V_0$, with

$$SV(x) = \sup_b [\tau B(x, b) + V(X(\tau, x, b))]. \tag{3.10}$$

Especially for the application to continuous-time models, it is important to have estimates of convergence uniform in $n = t/\tau$ for bounded total time $t = n\tau$.

As a preliminary step, let us prove a rather standard fact about the propagation of continuity by the operator S.

Proposition 3.1.1. *Let V, B, Q be bounded continuous functions that are Lipschitz continuous in x, with*

$$\varkappa_V = \|V\|_{Lip}, \quad \varkappa_B = \sup_b \|B(., b)\|_{Lip} < \infty, \quad \omega = \sup_b \|Q(., b)\|_{bLip} < \infty. \tag{3.11}$$

Then $S^n V \in C_{bLip}$ for all n and

$$\|S^n V\| \le t\|B\| + \|V\|, \quad \|S^n V\|_{Lip} \le (t\varkappa_B + \varkappa_V)e^{t\omega} \tag{3.12}$$

for $t = n\tau$.

Proof. The first equation in (3.12) follows from the definition of SV. Next, by (2.45) and (2.11),

$$|X(\tau, x, b) - X(\tau, y, b)| \le |x - y|e^{t\omega},$$

and therefore $|SV(x) - SV(y)|$ does not exceed

$$\sup_b [\tau B(x, b) + V(X(\tau, x, b)) - \tau B(y, b) - V(X(\tau, y, b))] \le |x - y|(\tau \varkappa_B + \varkappa_V e^{t\omega}).$$

Similarly,

$$|S^2 V(x) - S^2 V(y)| \le \sup_b [\tau B(x, b) + SV(X(\tau, x, b)) - \tau B(y, b) - SV(X(\tau, y, b))]$$

$$\leq |x - y|(\tau \varkappa_B + \|SV\|_{Lip} e^{t\omega}) \leq |x - y|(\tau \varkappa_B + (\tau \varkappa_B + \varkappa_V e^{t\omega})e^{t\omega}).$$

By induction, we obtain that

$$\|S^n V\|_{Lip} \leq \tau \varkappa_B (1 + e^{\tau \omega} + \cdots + e^{\tau \omega (n-1)}) + \varkappa_V e^{\tau \omega n} \leq n \tau \varkappa_B e^{\tau \omega n} + \varkappa_V e^{\tau \omega n},$$

implying the second estimate in (3.12). $\qquad\qquad\qquad\qquad\qquad\qquad\qquad\square$

Theorem 3.1.1. *(i) Assume that (3.11) and (3.6) hold. Then for all $\tau \in (0, 1], n \in N$, and $t = \tau n$,*

$$\|S^n[N]V - S^n V\| \leq C(d, \omega)(t \varkappa_B + \varkappa_V)e^{t\omega}(t\sqrt{1/(\tau N)} + |x(N) - x|) \quad (3.13)$$

with a constant $C(d, \omega)$ depending on d and ω. In particular, for $\tau = N^{-\epsilon}$ with $\epsilon \in (0, 1)$, this turns into

$$\|S^n[N]V - S^n V\| \leq C(d, \omega)(t \varkappa_B + \varkappa_V)e^{t\omega}(t N^{-(1-\epsilon)/2} + |x(N) - x|). \quad (3.14)$$

(ii) If there exists a Lipschitz continuous optimal policy $\pi = \{b_k\}$, $k = 1, \cdots, n$, for the limiting optimization problem, then π is approximately optimal for the N-agent problem, in the sense that for every $\epsilon > 0$, there exists N_0 such that for all $N > N_0$,

$$|V_n^N(x(N)) - V_n^{N,\pi}(x(N))| \leq \epsilon.$$

Proof. (i) Let $L = \sup_{k \leq n} \|S^n V\|_{Lip}$. By (3.12) it is bounded by $(t \varkappa_B + \varkappa_V)e^{t\omega}$. By (2.58),

$$|S[N]V(x) - SV(x)| \leq \sup_b |\mathbf{E}V(X_x^N(\tau)) - V(X_x(\tau))| \leq C(d, \omega)L\sqrt{\tau/N},$$

where we have used the condition $\tau \leq 1$ to estimate $e^{3\tau\omega} \leq C(\omega)$. Next,

$$|S^2[N]V(x) - S^2 V(x)| \leq \sup_b |\mathbf{E}S[N]V(X_x^N(\tau)) - SV(X_x(\tau))|$$

$$\leq \sup_b \mathbf{E}|S[N]V(X_x^N(\tau)) - SV(X_x^N(\tau))| + \sup_b |\mathbf{E}SV(X_x^N(\tau)) - SV(X_x(\tau))|$$

$$\leq C(d, \omega)L\sqrt{\tau/N} + C(d, \omega)L\sqrt{\tau/N} \leq 2C(d, \omega)L\sqrt{\tau/N}.$$

It follows by induction that

$$\|S^n[N]V - S^n V\| \leq C(d, \omega)n L\sqrt{\tau/N} = C(d, \omega)t L\sqrt{1/(\tau N)}, \quad (3.15)$$

yielding (3.13).

(ii) One shows as above that for every Lipschitz continuous policy π, the corresponding value functions $V^{\pi,N}$ converge. Combined with (i), this yields statement (ii). $\qquad\qquad\qquad\qquad\qquad\qquad\qquad\qquad\qquad\qquad\qquad\square$

3.2 Infinite Horizon: Discounted Payoff

The standard optimization problem of infinite horizon planning related to the finite horizon problem of optimizing (3.2) is the problem of maximizing the discounted sum

$$\Pi^{\pi,N}(x(N)) = \mathbf{E}_{N,x(N)} \sum_{k=0}^{\infty} \tau \beta^k B(x_k, b_k), \tag{3.16}$$

with $\beta \in (0, 1)$, where as above,

$$x_k = X_N(\tau, x_{k-1}, b_{k-1}), \quad k = 1, 2, \cdots,$$

$\pi = \{b_k\}$ and $b_k = b_k(x_k)$ are measurable functions depending on the current state $x = x_k$.

In the law of large numbers limit $N \to \infty$, we expect the limiting problem for the principal to be the maximization of the reward

$$\Pi^{\pi}(x) = \sum_{k=0}^{\infty} \beta^k \tau B(x_k, b_k) \tag{3.17}$$

with $x_k = X(\tau, x_{k-1}, b_{k-1})$.

Notice first that the solution to the finite-time discounting problem of the maximization of the payoff

$$V_n^{\pi,N}(x(N)) = \mathbf{E}_{N,x(N)} \left[\tau B(x_0, b_0) + \cdots + \beta^{n-1} \tau B(x_{n-1}, b_{n-1}) + \beta^n V(x_n) \right] \tag{3.18}$$

is given by the iterations

$$V_n^N = S_{\beta}[N] V_{n-1}^N = S_{\beta}^n[N] V \tag{3.19}$$

of the corresponding discounted Shapley operator

$$S_{\beta}[N] V(x) = \sup_b \left[\tau B(x, b) + \beta \mathbf{E} V(X_N(\tau, x, b)) \right]. \tag{3.20}$$

Similarly, the solution to the corresponding limiting discounted problem

$$V_n(x) = \max_{\pi} \left[\tau B(x_0, b_0) + \cdots + \beta^{n-1} \tau B(x_{n-1}, b_{n-1}) + \beta^n V(x_n) \right] \tag{3.21}$$

is given by the iterations

$$V_n = S_{\beta} V_{n-1}^N = S_{\beta}^n V \tag{3.22}$$

of the corresponding discounted Shapley operator

$$S_\beta V(x) = \sup_b \left[\tau B(x, b) + \beta V(X(\tau, x, b)) \right].$$ (3.23)

As a preliminary step, let us recall a standard fact about the finite-step approximations to the optimal Π^π.

Proposition 3.2.1. *Assume*

$$\omega = \sup_b \| Q(., b) \|_{bLip} < \infty.$$

Let B and V be bounded continuous functions. Then the sequence $S_\beta^n V(x)$ converges, as $n \to \infty$, to the discounted infinite horizon optimal reward

$$\Pi(x) = \sup_\pi \Pi^\pi(x),$$

and the sequence $S_\beta^n[N]V(x(N))$ converges, as $n \to \infty$, to the discounted infinite horizon optimal reward

$$\Pi^N(x(N)) = \sup_\pi \Pi^{\pi, N}(x(N)).$$

Proof. Since

$$\| S_\beta^n V - S_\beta^n \tilde{V} \| \le \beta^n \| V - \tilde{V} \|,$$

$$\| S_\beta^n[N]V - S_\beta^n[N]\tilde{V} \| \le \beta^n \| V - \tilde{V} \|,$$

it follows that if the sequence $S_\beta^n V$ or $S_\beta^n[N]V$ converges for some V, then these iterations converge to the same limit for every bounded V. But for $V = 0$, we see directly that

$$\| S_\beta^n V - \Pi \| \le \beta^n \| B \| \frac{2}{1 - \beta},$$

$$\| S_\beta^n[N]V - \Pi^N \| \le \beta^n \| B \| \frac{2}{1 - \beta}.$$ $\square$

Theorem 3.2.1. *Assume that* (3.11) *and* (3.6) *hold and let*

$$\beta e^{\tau \omega} \le \beta_0 < 1.$$

Then the discounted optimal rewards

$$\Pi^N(x(N)) = \sup_\pi \Pi^{\pi, N}(x(N))$$

converge, as $N \to \infty$ and $x(N) \to x$, to the discounted best reward

$$\Pi(x) = \sup_{\pi} \Pi^{\pi}(x).$$

Proof. Following the arguments of Proposition 3.1.1, we obtain

$$\|S_{\beta}^n V\| \le \tau \|B\| \frac{1}{1 - \beta_0} + \beta_0^n \|V\|,$$

$$\|S_{\beta}^n V\|_{Lip} \le \tau \varkappa_B (1 + \beta e^{\tau\omega} + \cdots + \beta^{n-1} e^{\tau\omega(n-1)}) + \beta^n \varkappa_V e^{\tau\omega n} \le \tau \varkappa_B \frac{1}{1 - \beta_0} + \beta_0^n \varkappa_V,$$

that is, in contrast to optimization without discounting, these norms are bound uniformly in the number of steps used.

Estimating the differences $\|S_{\beta}^n[N]V - S_{\beta}^n V\|$ as in the proof of Theorem 3.1.1 yields

$$\|S_{\beta}^n[N]V - S_{\beta}^n V\| \le C \frac{\sqrt{\tau}}{\sqrt{N}} (\beta \|S_{\beta}^n V\|_{bLip} + \cdots + \beta^n \|S_{\beta} V\|_{bLip})$$

$$\le C \frac{\sqrt{\tau}}{\sqrt{N}} \left(\tau \|B\|_{bLip} \frac{1}{(1 - \beta_0)^2} + \beta_0^{n+1} \|V\|_{bLip} \right).$$

Since $S_{\beta}^n V(x(N))$ converges to $\Pi(x)$ as $n \to \infty$, it follows that $S_{\beta}^n[N]V(x)$ converges to $\Pi(x)$ as $n \to \infty$ and $N \to \infty$. □

Remark 14. *The optimal payoffs $\Pi(x)$ and $\Pi^N(x)$ are the fixed points of the Shapley operator: $S_{\beta}\Pi = \Pi$, $S_{\beta}[N]\Pi^N = \Pi$. This fact can be used as a basis for another proof of Theorem 3.2.1.*

3.3 Continuous-Time Modeling

Here we initiate the analysis of the optimization problem for a forward-looking principal. Namely, let the state space of the group again be given by vectors $x = (n_1, \cdots, n_d)/N$ from the lattice $\mathbf{Z}_+^d/N$, but the efforts (budget) b of the major player are chosen continuously in time, aiming at optimizing the payoff

$$\int_t^T B(x(s), b(s)) \, ds + S_T(x(T)),$$

where B, S_T are some continuous functions uniformly Lipschitz in all their variables. The optimal payoff of the major player is thus

$$S_N(t, x(N)) = \sup_{b(.)} \mathbf{E}^N_{x(N)} \left\{ \int_t^T B(x(s), b(s)) \, ds + S_T(x(T)) \right\}, \tag{3.24}$$

where $\mathbf{E}^N_x$ is the expectation of the corresponding Markov process starting at the position (x) at time t, and $\tilde{U}$ is some class of controls (say, piecewise constant). We are now in the standard Markov decision setting of a controlled Markov process generated by the operator $L_{b,N}$ from (3.1).

As was shown above, the operators $L_{b,N}$ tend to a simple first-order PDO, so that the limiting optimization problem of the major player turns out to be the problem of finding

$$S(t, x) = \sup_{b(.)} \left\{ \int_t^T B(x(s), b(s)) ds + S_T(x(T)) \right\}, \tag{3.25}$$

where $x(s)$ solve the system of equations

$$\dot{x}_j = \sum_i x_i Q_{ij}(x, b), \quad j = 1, ..., d. \tag{3.26}$$

The well-posedness of this system is a straightforward extension of the well-posedness of equations (2.41).

It is well known (see, e.g., [147] or any textbook on deterministic optimal control) that the optimal payoff $S(t, x)$ of (3.25) represents the unique generalized solution (so-called viscosity solution) to the HJB equation

$$\frac{\partial S}{\partial t} + \sup_b \left[B(x, b) + \left(\frac{\partial S}{\partial x}, x Q(x, b) \right) \right] = 0, \tag{3.27}$$

with the initial (or terminal) condition $S(T, x) = S_T(x)$.

Instead of proving the convergence $S_N(t, x(N)) \to S(t, x)$, we shall concentrate on a more practical issue, comparing the corresponding discrete-time approximations, since these approximations are usually exploited for practical calculations of S_N or S.

The discrete-time approximation to the limiting problem of finding (3.25) is the problem of finding

$$V_{t,n}(x) = \sup_\pi V^\pi_{t,n}(x) = \sup_\pi \left[\tau B(x_0, b_0) + \cdots + \tau B(x_{n-1}, b_{n-1}) + V(x_n) \right], \tag{3.28}$$

where $\tau = (T - t)/n$, $x_0 = x$, $V(x) = S_T(x)$, and

$$x_k = X(\tau, x_{k-1}, b_{k-1}), \quad k = 1, 2, \cdots, \tag{3.29}$$

with $X(t, x, b)$ solving equation (3.8) with the initial condition x at time $t = 0$. It is known (see, e.g., Theorem 3.4 of [147]) that the discrete approximations $V_n(x)$ approach the optimal solution $S(T - t, x)$ given by (3.25) and solving the Cauchy problem for the HJB (3.27).

The discrete-time approximation to the initial optimization problem is the problem of finding

$$V_{t,n}^N(x_0) = \sup_\pi V_{t,n}^{\pi,N}(x_0)$$

$$= \sup_\pi \mathbf{E}_{N,x(N)} \left[\tau B(x_0, b_0) + \cdots + \tau B(x_{n-1}, b_{n-1}) + V_0(x_n) \right], \tag{3.30}$$

where $x_k = X_N(\tau, x_{k-1}, b_{k-1})$, with $X_N(t, x, b)$ denoting the Markov process with generator (3.1) and with the strategies $\pi = \{b_k\}$ as in Section 3.1.

Remark 15. *It is also known, see e.g., Theorem 4.1 of [86], that $V_n^N(x)$ with $V_0 = S_T$ approach the optimal solutions $S_N(T - t, x)$ given by (3.24) and solving a certain HJB equation.*

Theorem 3.3.1. *Let B, S_T be uniformly Lipschitz in all their variables. Then for all $x, t \in [0, T]$, and $\tau = N^{-\epsilon}$ with $\epsilon \in (0, 1)$,*

$$|V_{t,n}^N(x) - V_{t,n}(x)| \le C(d, \omega, T)(\varkappa_B + \varkappa_V)(N^{-(1-\epsilon)/2} + |x(N) - x|). \tag{3.31}$$

And consequently, $V_{t,n}^N(x)$ converge as $N \to \infty$ (and $n = (T - t)/\tau$, $\tau = N^{-\epsilon}$) to the optimal solution $S(T - t, x)$ given by (3.25) and solving the Cauchy problem for the HJB (3.27).

Proof. This is a direct consequence of Theorem 3.1.1 (i). □

One can also imagine a situation in which changing the budget bears some costs, so that instantaneous adjustments of policies become infeasible, in which case the efforts (budget) b of the major player can evolve depending on some more flexible control parameter $u \in U \in \mathbf{R}^r$, for instance, according to the equation $\dot{b} = u$. In this case, the payoff of the major player can be given, as usual, by the function

$$\int_t^T J(x(s), b(s), u(s))\, ds + S_T(x(T), b(T)),$$

where J, S_T are some Lipschitz continuous functions. The optimal payoff of the major player is then

$$S_N(t, x(N), b) = \sup_{u(.)} \mathbf{E}_{x(N),b}^N \left\{ \int_t^T J(x(s), b(s), u(s))\, ds + S_T(x(T), (b(T)) \right\}, \tag{3.32}$$

where $\mathbf{E}_{x,b}^N$ is the expectation of the corresponding Markov process starting at the position (x, b) at time t, and the corresponding limiting optimal payoff is

$$S(t, x, b) = \sup_{u(.)} \left\{ \int_t^T J(x(s), b(s), u(s)) ds + S_T(x(T), (b(T)) \right\}, \tag{3.33}$$

where $(x(s), (b(s))$ (depending on $u(.)$) solve the system of equations

$$\dot{b} = u, \quad \dot{x}_j = \sum_i x_i Q_{ij}(x, b), \quad j = 1, ..., d.$$

The well-posedness of this system is a straightforward extension of the well-posedness of equations (2.41). Everything extends directly to this case, where the limiting HJB equation becomes

$$\frac{\partial S}{\partial t} + \sup_u \left[J(x, b, u) + \left(\frac{\partial S}{\partial x}, x Q(x, b) \right) + u \frac{\partial S}{\partial b} \right] = 0, \qquad (3.34)$$

with the initial (or terminal) condition $S(T, x, b) = S_T(x, b)$.

A similar analysis can be performed for the continuous-time discounted payoff setting, the limiting payoff being then given by a solution to the corresponding stationary HJB equation.

Remark 16. *Performing the same analysis for the case of the nonexponential jumps of Section 2.14 would lead to a limit that would not be controlled by the HJB equation (3.27), but its fractional counterpart, the fractional HJB, derived in the general setting in [150] and [151], which in the present case gets the form*

$$D_{T-*}^{A(x)} S(x, s) = \sup_b \left[B(x, b) + \left(\frac{\partial S}{\partial x}, x Q(x, b) \right) \right], \quad s \in [0, T]. \qquad (3.35)$$

3.4 Several Major Players

Let us extend the above result to the case of several players competing in the background of a pool of small players.

As was pointed out in Chapter 1, in games with several principals, one can naturally distinguish two cases: the pool of small players is common for the major agents (think about advertising models, or several states together fighting a terrorist group) or is specific to each major agent (think about generals controlling armies, big banks controlling their subsidiaries, etc.). Of course mathematically, the second case can be embedded in the first case (with an appropriate increase of dimension). However, the second case is quite special, because in the deterministic limit (with which we are mostly concerned), it turns into the differential games with so-called *separated dynamics*, whereby each player controls its own position independently of other players (as in the classical games of pursuit and evasion). And this case is much better developed in the theory of games.

Let us start with the case of two major players, I and II, playing a zero-sum game. We shall first discuss the general case of a common pool of small players and then point out the specific features of "separated dynamics."

Let b^1 and b^2 denote the control parameters of players I and II respectively, each taken from a compact subset of a Euclidean space. Let player I try to maximize the reward $B(x, b^1, b^2, N)$ and let player II try to minimize it.

The corresponding modification of Section 3.1 leads to the study of the Markov chains $X_N(t, x, b^1, b^2)$, which are the chains generated by operators (3.1) with b substituted by the pair (b^1, b^2).

Starting again with the discrete setting, we assume that players I and II update their strategies in discrete times $\{k\tau\}$, $k = 0, 1, \cdots, n - 1$. The rules of the game are distinguished by the order of the moves of players I and II.

In the case of the *upper game*, we assume that player I always makes the first move, which directly becomes known to player II. This leads to the problem of finding

$$V_{n,up}^N(x(N)) = \inf_{\pi^1} \sup_{\pi^2} \mathbf{E}_{N,x(N)} \left[\tau B(x_0, b_0^1, b_0^2) + \cdots + \tau B(x_{n-1}, b_{n-1}^1, b_{n-1}^2) + V(x_n) \right],$$
(3.36)

with $x_0 = x(N) \in \mathbf{Z}_+^d / N$. Here strategies π^1 are rules that assign the controls b_k^1 of player I on the basis of the history of the chain up to steps k, that is, on the basis of known $x_0, \cdots, x_k, b_0^1, \cdots, b_{k-1}^1, b_0^1, \cdots, b_{k-1}^1$, strategies π^2 are rules that assign the controls b_k^2 of player II on the basis of the history of the chain up to steps k plus the choice b_k^1, that is, on the basis of known $x_0, \cdots, x_k, b_0^1, \cdots, b_{k-1}^1, b^k, b_0^1, \cdots, b_{k-1}^1$, and

$$x_k = X_N(\tau, x_{k-1}, b_{k-1}^1, b_{k-1}^2), \quad k = 1, 2, \cdots.$$

Similarly, in the case of the *lower game*, we assume that player II always makes the first move, which directly becomes known to player I. This leads to the problem of finding

$$V_{n,low}^N(x(N)) = \sup_{\pi^2} \inf_{\pi^1} \mathbf{E}_{N,x(N)} \left[\tau B(x_0, b_0^1, b_0^2) + \cdots + \tau B(x_{n-1}, b_{n-1}^1, b_{n-1}^2) + V(x_n) \right].$$
(3.37)

Here π^2 are strategies assigning controls b_k^2 of player II on the basis of the history of the chain up to steps k, and π^1 are strategies that assign controls b_k^1 of player I on the basis of the history of the chain up to steps k plus the choice b_k^2.

The values $V_{n,up}^N$ and $V_{n,low}^N$ are called the *upper* and *lower values* of the multistage stochastic game with payoff $B(x, b^1, b^2, N)$ and Markov dynamics $X_N(\tau, x, b^1, b^2)$.

It is a standard fact (see, e.g., [194] or [144]) that $V_{n,up} \geq V_{n,low}$, and each of these values can be calculated as the iterations of certain Shapley operators:

$$V_{n,low}^N = S_{low}[N]V_{n-1,low}^N = S_{low}^n[N]V, \quad V_{n,up}^N = S_{up}[N]V_{n-1,up}^N = S_{up}^n[N]V,$$
(3.38)

where

$$S_{low}[N]V(x) = \sup_{b^2} \inf_{b^1} \left[\tau B(x, b_1, b_2, N) + \mathbf{E}V(X_N(\tau, x, b_1, b_2)) \right], \quad (3.39)$$

$$S_{up}[N]V(x) = \inf_{b^1} \sup_{b^2} \left[\tau B(x, b_1, b_2, N) + \mathbf{E}V(X_N(\tau, x, b_1, b_2)) \right]. \quad (3.40)$$

We are again interested in the law of large numbers limit, $N \to \infty$, where we expect the limiting values $V_{n,up}$ and $V_{n,low}$ to be given by the corresponding problems with the limiting deterministic dynamics:

$$V_{n,up}(x) = \inf_{\pi^1} \sup_{\pi^2} \left[\tau B(x_0, b_0^1, b_0^2) + \cdots + \tau B(x_{n-1}, b_{n-1}^1, b_{n-1}^2) + V(x_n) \right],$$
(3.41)

$$V_{n,low}(x) = \sup_{\pi^2} \inf_{\pi^1} \left[\tau B(x_0, b_0^1, b_0^2) + \cdots + \tau B(x_{n-1}, b_{n-1}^1, b_{n-1}^2) + V(x_n) \right],$$
(3.42)

with $x_0 = x$, the same rules for strategies as above, but with the dynamics

$$x_k = X(\tau, x_{k-1}, b_{k-1}^1, b_{k-1}^2), \quad k = 1, 2, \cdots,$$

where $X(t, x, b^1, b^2)$ denotes the solution of the kinetic equation

$$\dot{x}_k = \sum_{i=1}^{d} x_i Q_{ik}(x, b^1, b^2), \quad k = 1, ..., d,$$
(3.43)

with the initial data x at time zero.

As in the case of finite N games, the values $V_{n,up}$ and $V_{n,low}$ are given by the iterations of the corresponding Shapley operators:

$$V_{n,low} = S_{low} V_{n-1,low} = S_{low}^n V, \quad V_{n,up} = S_{up} V_{n-1,up} = S_{up}^n V,$$
(3.44)

where

$$S_{low} V(x) = \sup_{b^2} \inf_{b^1} \left[\tau B(x, b_1, b_2) + V(X(\tau, x, b_1, b_2)) \right],$$
(3.45)

$$S_{low} V(x) = \inf_{b^1} \sup_{b^2} \left[\tau B(x, b_1, b_2) + V(X(\tau, x, b_1, b_2)) \right].$$
(3.46)

The following analogue of Theorem 3.1.1 is straightforward.

Theorem 3.4.1. *(i) Assume*

$$\varkappa_V = \|V\|_{Lip}, \quad \varkappa_B = \sup_{b^1, b^2} \|B(., b^1, b^2)\|_{Lip} < \infty, \quad \omega = \sup_{b^1, b^2} \|Q(., b^1, b^2)\|_{bLip} < \infty,$$
(3.47)

and (3.6). *Then for all* $\tau \in (0, 1]$, $n \in N$, *and* $t = \tau n$,

$$\max(\|S_{low}^n[N]V - S_{low}^n V\|, \|S_{up}^n[N]V - S_{up}^n V\|)$$

$$\leq C(d, \omega)(t \varkappa_B + \varkappa_V) e^{t\omega} (t\sqrt{1/(\tau N)} + |x(N) - x|)$$
(3.48)

with a constant $C(d, \omega)$ *depending on* d *and* ω. *In particular, for* $\tau = N^{-\epsilon}$ *with* $\epsilon \in (0, 1)$, *this becomes*

$$\max(\|S_{low}^n[N]V - S_{low}^n V\|, \|S_{up}^n[N]V - S_{up}^n V\|)$$

$$\leq C(d, \omega)(t \varkappa_B + \varkappa_V)e^{t\omega}(t N^{-(1-\epsilon)/2} + |x(N) - x|). \tag{3.49}$$

(ii) If there exist Lipschitz continuous optimal policies π^1, π^2 *for the lower or upper game (the policies giving a minimax in (3.41) or maximin in (3.42) respectively, then* π^1, π^2 *are approximately optimal for the N-agent problem.*

In the continuous-time setting, the efforts b^1, b^2 of the major players are chosen continuously, with player I aiming at maximizing and player II at minimizing the payoff

$$\int_t^T B(x(s), b^1(s), b^2(s)) \, ds + S_T(x(T)), \tag{3.50}$$

where J, S_T are some continuous functions uniformly Lipschitz in all their variables.

The discrete-time approximation to this game is the problem of finding the upper and lower values (3.36), (3.37) with the limiting values (3.41), (3.42).

It is known (see, e.g., [86], [194]) that under the assumptions of Lipschitz continuity of all functions involved, the limit of values (3.41) and (3.42) as $n \to \infty$, $\tau = t/n$, are the generalized (so-called *viscosity*) solutions to the upper and lower HJB–Isaacs equations, respectively:

$$\frac{\partial S}{\partial t} + \inf_{b^1} \sup_{b^2} \left[B(x, b^1, b^2) + \left(\frac{\partial S}{\partial x}, x Q(x, b^1, b^2) \right) \right] = 0, \tag{3.51}$$

$$\frac{\partial S}{\partial t} + \sup_{b^2} \inf_{b^1} \left[B(x, b^1, b^2) + \left(\frac{\partial S}{\partial x}, x Q(x, b^1, b^2) \right) \right] = 0, \tag{3.52}$$

with the terminal condition S_T.

In many cases (see, e.g., [86], [194]), one can show that solutions to (3.51) and (3.52) coincide, the common value $V(t, x)$ being known as the *value* of the zero-sum differential game with the dynamics

$$\dot{x}_k = \sum_{i=1}^d x_i Q_{ik}(x, b^1, b^2), \quad k = 1, ..., d, \tag{3.53}$$

and payoff function (3.50). Such differential games are said to *have a value*.

The following result is a straightforward extension of Theorem 3.3.1.

Theorem 3.4.2. *Assume that the game with dynamics* (3.53) *and payoff* (3.50) *has a value* $V(t, x)$ *and that all functions involved as parameters are Lipschitz continuous. Then for all* x, $t \in [0, T]$, *and* $\tau = N^{-\epsilon}$ *with* $\epsilon \in (0, 1)$, *both the upper and lower values* (3.36), (3.37) *converge to* $V(t, x)$ *as* $N \to \infty$.

Let us describe in more detail the setting of two players controlling different pools of minor players. Suppose we are given two families of Q-matrices $\{Q(t, u, x) = (Q_{ij})(u, x)\}$ and $\{P(t, v, x) = (P_{ij})(v, x)\}$, $i, j = 1, \cdots d$, depending on $x \in \Sigma_d$ and parameters u and v from two subsets U and V of Euclidean spaces. Any two given bounded measurable curves $u(t)$, $v(t)$, $t \in [0, T]$, define a Markov chain on

$$(\Sigma_d \cap \mathbf{Z}_+^d / N) \times (\Sigma_d \cap \mathbf{Z}_+^d / M),$$

specified by the generator

$$L_{t,u(t),v(t)}^{N,M} f(x, y) = N \sum_{i=1}^{d} \sum_{j=1}^{d} x_i Q_{ij}(t, u(t), x)[f(x - e_i/N + e_j/N, y) - f(x, y)]$$

$$+ M \sum_{i=1}^{d} \sum_{j=1}^{d} y_i P_{ij}(t, v(t), y)[f(x, y - e_i/M + e_j/M) - f(x, y)]. \tag{3.54}$$

For $f \in C^1(\Sigma_d \times \Sigma_d)$,

$$\lim_{N,M \to \infty} L_{t,u(t),v(t)}^{N,M} f(x, y) = \Lambda_{t,u(t),v(t)} f(x, y),$$

where

$$\Lambda_{t,u(t),v(t)} f(x, y) = \sum_{k=1}^{d} \sum_{i \neq k} [x_i Q_{ik}(t, u(t), x) - x_k Q_{ki}(t, u(t), x)] \frac{\partial f}{\partial x_k}(x, y)$$

$$+ \sum_{k=1}^{d} \sum_{i \neq k} [y_i P_{ik}(t, v(t), y) - y_k P_{ki}(t, v(t), y)] \frac{\partial f}{\partial y_k}(x, y). \tag{3.55}$$

The corresponding controlled characteristics are governed by the equations

$$\dot{x}_k = \sum_{i \neq k} [x_i Q_{ik}(t, u(t), x) - x_k Q_{ki}(t, u(t), x)] = \sum_{i=1}^{d} x_i Q_{ik}(t, u(t), x), \quad k = 1, ..., d,$$

$$\tag{3.56}$$

$$\dot{y}_k = \sum_{i \neq k} [y_i P_{ik}(t, v(t), y) - y_k P_{ki}(t, v(t), y)] = \sum_{i=1}^{d} y_i P_{ik}(t, v(t), y), \quad k = 1, ..., d.$$

$$\tag{3.57}$$

For a given $T > 0$, let us denote by $\Gamma(T, \tau)^N$ the discrete-time stochastic game with the dynamics specified by the generator (3.54), with the objective of the player I (controlling Q via u) being to maximize the payoff $\mathbf{E}V_T(x(T), y(T))$ for a given function V_T (terminal payoff), and with the objective of player II (controlling P via v) being to minimize this payoff (zero-sum game), while the decisions for choosing u and v are taken at time τk, $k \in \mathbf{N}$. As previously, we want to approximate it by the deterministic zero-sum differential game $\Gamma(T)$, defined by the dynamics (3.56), (3.57) and the payoff of player I, given by $V_T(X_{t,x}(T), Y_{t,y}(T))$.

For this game with the separated dynamics it is known that the value $V(t, x, y)$ exists (see, e.g., [86] or [169]). Hence Theorem 3.4.2 applies, so that the corresponding upper and lower discrete-time values converge to $V(t, x)$ as as $N \to \infty$, where V is the generalized (viscosity or minimax) solution (see [86] or [218]) of the HJB–Isaacs equation

$$\frac{\partial S}{\partial t} + \sup_u \left(\frac{\partial S}{\partial x}, x Q(x, u) \right) + \inf_v \left(\frac{\partial S}{\partial y}, y P(y, v) \right) = 0. \qquad (3.58)$$

One can similarly explore the corresponding competitive version of the setting (3.32), (3.33) that takes into account the costs of changing the budgets of the major players.

The situation with many major players is quite similar. Though the results on the Nash equilibria in several-player differential games are well less developed (see, however, [169], [194]), the general conclusion from the above discussion is as follows: *whenever the discrete approximations to a certain differential game have a common limit, the same common limit exists for the discrete approximations to the corresponding stochastic game of the major players and N minor players as $N \to \infty$, $\tau = N^{-\epsilon}, \epsilon \in (0, 1)$.*

Remark 17. *The original paper [138] dealing with Markov chains generated by (3.54) has a gap in the argument showing directly the convergence of the limiting $\tau \to 0$ dynamics of a fixed number of players to the limiting differential game. This gap is corrected by showing the convergence of the discrete approximations $\tau \to 0$ with $\tau \sim N^{-\epsilon} \sim M^{-\epsilon}$.*

3.5 Turnpike Theory for Nonlinear Markov Games

In Section 3.2, we discussed infinite-horizon problems with discounting payoff. Here we shall touch upon the analysis of long-time optimization problems without discounting, working in the framework of Section 3.3. This analysis also allows one to link the best-response principal modeling of Chapter 2 with the present setting of a forward-looking principal.

For simplicity, let us assume, as in Section 2.6, that for every x there exists a unique point of maximum $b^*(x)$ of the function $B(x, b)$.

Our analysis will be based on the following assumption (A) on the existence of an optimal steady state for the kinetic dynamics:

(A) There exists a unique x^* such that

$$0 = B(x^*, b^*(x^*)) = \max_x B(x, b^*(x)) = \max_{x,b} B(x, b). \tag{3.59}$$

Moreover, $(x^*, b^*(x^*))$ is an asymptotically stable stationary point of the kinetic equations, that is, $x^* Q(x^*, b^*(x^*)) = 0$, and there exists a neighborhood U of x^* such that for all $x \in U$, the solution $X_x(t)$ of the best-response dynamics (2.71), that is, of the equation $\dot{x} = x Q(x, b^*(x))$ starting at x, converges to x^* exponentially:

$$\|X_x(t) - x^*\| \le e^{-\lambda t} \|x - x^*\|. \tag{3.60}$$

Let us also assume a power-type nondegeneracy of the maximum point x^*, that is, the following condition:

(B) For x in U, J behaves like a power function:

$$J(x, b^*(x)) \sim -\tilde{J} |x - x^*|^\alpha \tag{3.61}$$

for some constants $\tilde{J}, \alpha > 0$.

Under these assumptions, the stationary point solution $X(t) = x^*$ of the best-response dynamics (2.71) represents a *turnpike* for the optimization problem (3.25)–(3.26) in the following sense.

Theorem 3.5.1. *Under assumptions (A) and (B), let $x \in U$, S_T be bounded, and let $\tilde{X}_x(s)$ be the optimal trajectory for the optimization problem (3.25)–(3.26). Then for every sufficiently small ϵ and time horizon $T - t$, the time τ that this trajectory spends away from the ϵ-neighborhood of x^* satisfies the following bound:*

$$\tau \le \frac{1}{\epsilon^\alpha} \left[|x - x^*|^\alpha \lambda^{-1} + (\sup S_T - S(x^*)) \tilde{J}^{-1} \right]. \tag{3.62}$$

Thus for large $T - t$, an optimal trajectory spends most of the time near x^*. Moreover, the payoff V_t for the optimization problem (3.25)–(3.26) has the following bounds:

$$C|x - x^*|^\alpha + S(x^*) \le V_t \le \sup S_T \tag{3.63}$$

with a constant C.

Proof. The cost on the trajectory $X_x(s)$ of the best-response dynamics (2.71) starting at x at time t is of order

$$S_T(x^*) - \int_t^T e^{-\lambda(s-t)} \tilde{J} |x - x^*|^\alpha \, ds = S_T(x^*) - \frac{\tilde{J}}{\lambda} |x - x^*|^\alpha (1 - e^{-(T-t)}). \tag{3.64}$$

If a trajectory spends time τ away from the ϵ-neighborhood of x^*, then the cost on this trajectory does not exceed

$$-\tau \tilde{J} \epsilon^\alpha + \sup S_T.$$

If it is an optimal trajectory, it should perform better than $X_x(s)$. Comparing with (3.64) yields (3.62). The lower bound in (3.63) is again obtained from the trajectory $X_x(s)$. $\qquad \square$

Combining Theorems 3.5.1 and 3.3.1, we obtain the following result.

Theorem 3.5.2. *Under the assumptions of Theorem 3.5.1, the optimal cost $V_{t,n}^N$ for the approximate Markov chain (from Theorem 3.3.1) is approximated by the cost on the limiting best-response dynamics $X_x(s)$.*

For a general introduction to the turnpike theory, related background, and bibliography, we refer the reader to the paper [152]; see also monograph [235] for the case of deterministic control.

Chapter 4
Models of Growth Under Pressure

The results of this chapter extend the results of Chapters 2 and 3 to the case of a countable state space of small players, and moreover, to the case of processes that allow for a change in the number of particles (thus going beyond the simple migrations that we have played with so far), where physical particles correspond in this setting to the coalitions (stable groups) of agents. This extension is carried out in order to include important models of evolutionary coalition-building, merging and splitting (banks, subsidiaries, etc.), strategically enhanced preferential attachment, and many others. The mathematics of this chapter is more demanding than in the rest of our presentation, and its results are not used in other parts of the book. It is based on some elements of infinite-dimensional analysis, the analysis of functions on the Banach space of sequences l_1 and of the ODEs in this space. We begin with brief description of the tools used.

4.1 Preliminaries: ODEs in l_1

We recall some elementary results on the positivity-preserving ODEs in l_1^+, starting with recalling some basic notation. As usual, we denote by $\mathbf{R}^\infty = \{(x_1, x_2, \cdots)\}$ the space of sequences of real numbers and by l_1 the Banach space of summable sequences

$$l_1 = \{(x_1, x_2, \cdots) : \|x\| = \sum_j |x_j| < \infty\}.$$

By $\mathbf{R}_+^\infty$ and l_1^+ we denote the subsets of these spaces with nonnegative coordinates x_j.

All notation for the basic classes of regular functions extends automatically to this infinite-dimensional setting. For instance, for a convex closed subset Z of l_1, $C(Z)$ denotes the space of bounded continuous functions equipped with the sup-norm: $\|f\| = \sup_x |f(x)|$. For these functions, the Lipschitz constant is

© Springer Nature Switzerland AG 2019
V. N. Kolokoltsov and O. A. Malafeyev, *Many Agent Games in Socio-economic Systems: Corruption, Inspection, Coalition Building, Network Growth, Security*, Springer Series in Operations Research and Financial Engineering, https://doi.org/10.1007/978-3-030-12371-0_4

$$\|f\|_{Lip} = \sup_{x \neq y} \frac{|f(x) - f(y)|}{\|x - y\|} = \sup_j \sup \frac{|f(x) - f(y)|}{|x_j - y_j|}, \tag{4.1}$$

where the last sup is the supremum over the pairs x, y that differ only in their jth coordinates. By $C_{bLip}(Z)$ we denote the space of bounded Lipschitz functions with the norm $\|f\|_{bLip} = \|f\| + \|f\|_{Lip}$. By $C^k(Z)$ we denote the space of functions on Z with continuous uniformly bounded partial derivatives equipped with the norm

$$\|f\|_{C^k(Z)} = \|f\| + \sum_{j=1}^{k} \|f^{(j)}(Z)\|,$$

where $\|f^{(j)}(Z)\|$ is the supremum of the magnitudes of all partial derivatives of f of order j. As in the case of functions on $\mathbf{R}^d$, we have $\|f\|_{C^1} = \|f\|_{bLip}$. The space $C(Z, l_1)$ of bounded continuous l_1-valued functions $f = (f_i) : Z \to l_1$ will usually also be denoted $C(Z)$, their norm being $\|f\| = \sup_{x \in Z} \|f(x)\|$.

We shall denote by (g, f) the inner product $(g, f) = \sum_{k=1}^{\infty} g_k f_k$ for the elements $g, f \in \mathbf{R}^\infty$ such that $\sum_{k=1}^{\infty} |g_k| |f_k| < \infty$.

As subsets Z of l_1 we shall mostly use the sets

$$\mathcal{M}(L) = \{y \in \mathbf{R}^\infty : \sum_j L_j |y_j| < \infty\}, \quad \mathcal{M}_{\leq \lambda}(L) = \{y \in \mathcal{M}(L) : \sum_j L_j |y_j| \leq \lambda\},$$

$$\mathcal{M}^+(L) = \mathcal{M}(L) \cap \mathbf{R}_+^\infty, \quad \mathcal{M}_{\leq \lambda}^+(L) = \mathcal{M}_{\leq \lambda}(L) \cap \mathbf{R}_+^\infty$$

for a nondecreasing sequence $L \in \mathbf{R}_+^\infty$ and $\lambda > 0$. For our purposes we will need only two examples L: $L(j) = 1$, for which $\mathcal{M}(L) = l_1$ and $L(j) = j$. If $L_j \to \infty$ as $j \to \infty$, then the sets $\mathcal{M}_{\leq \lambda}^+(L)$ are easily seen to be compact in l_1.

The function $g : \mathcal{M}^+(L) \to \mathcal{M}(L)$ for some $L \in \mathbf{R}_+^\infty$ is called *conditionally positive* if $g_j(x) \geq 0$ whenever $x_j = 0$. We will work with the l_1-valued ODEs

$$\dot{x} = g(x) \iff \{\dot{x}_j = g_j(x) \text{ for all } j = 1, \cdots\}, \tag{4.2}$$

with such a r.h.s. g. Since g may be defined only on $\mathcal{M}^+(L)$, we are thus reduced to the analysis of solutions that belong to $\mathcal{M}^+(L)$ for all times. The vector L (or the corresponding function on $\mathbf{N}$) is said to be a *Lyapunov vector* (or a *Lyapunov function*) for equation (4.2) (or for its r.h.s. function g) if the *Lyapunov condition*

$$(L, g(x)) \leq a(L, x) + b, \quad x \in \mathcal{M}^+(L), \tag{4.3}$$

holds with some constants a, b.

The Lyapunov function L is called *subcritical* (respectively *critical*) for g, or g is said to be *L-subcritical* (respectively *L- L-critical*) if $(L, g(x)) \leq 0$ (respectively $(L, g(x)) = 0$) for all $x \in \mathcal{M}^+(L)$. In the last case, L is also referred to as the *conservation law*.

Theorem 4.1.1. *Let* $g : \mathcal{M}^+(L) \to \mathcal{M}(L)$ *be conditionally positive and Lipschitz continuous on any set* $\mathcal{M}^+_{\leq\lambda}(L)$ *with some Lipschitz constant* $\varkappa(\lambda)$, *and let* L *be the Lyapunov vector for* g. *Then for all* $x \in \mathcal{M}^+(L)$, *there exists a unique global solution* $X(t, x) \in \mathcal{M}^+(L)$ *(defined for all* $t \geq 0$) *of equation* (4.2) *in* l_1 *with the initial condition* x. *Moreover, if* $x \in \mathcal{M}^+_{\leq\lambda}(L)$ *and* $a \neq 0$, *then*

$$X(t, x) \in \mathcal{M}^+_{\leq\lambda(t)}(L), \quad \lambda(t) = e^{at}\left(\lambda + \frac{b}{a}\right) - \frac{b}{a} = e^{at}\lambda + (e^{at} - 1)\frac{b}{a}. \quad (4.4)$$

If $a = 0$, *then the same holds with* $\lambda(t) = \lambda + bt$, *and if* f *is* L-*critical, then it holds with* $\lambda(t) = \lambda$. *Finally,* $X(t, x)$ *is Lipschitz as a function of* x:

$$\|X(t, x) - X(t, y)\| \leq e^{\varkappa(\lambda(t))}\|x - y\|. \quad (4.5)$$

Remark 18. *Intuitively, this result is clear. In fact, by conditional positivity, the vector field* $g(x)$ *on any boundary point of* $\mathbf{R}^\infty_+$ *is directed inside or tangent to the boundary, thus not allowing a solution to leave it. On the other hand, by the Lyapunov condition,*

$$(L, X(t, x)) \leq (L, x) + a\int_0^t (L, X(s, x))\, ds + bt,$$

implying (4.4) *by Grönwall's lemma. However, a rigorous proof is not fully straightforward, because the existence of a solution is not even clear: an attempt to construct it via a usual fixed-point argument reducing it to the integral equation* $x(t) = \int_0^t g(x(s))\, ds$ *or by the standard Euler or Peano approximations encounters a problem, since all these approximations may not preserve positivity (and thus may jump out of the domain on which* g *is defined).*

Proof. Assuming $a \neq 0$ for definiteness, let

$$M^x_{a,b}(t) = \left\{y \in \mathbf{R}^\infty_+ : (L, y) \leq e^{at}\left((L, x) + \frac{b}{a}\right) - \frac{b}{a}\right\}.$$

Fixing T, let us define the space $C_{a,b}(T)$ of continuous functions $y : [0, T] \mapsto \mathbf{R}^\infty_+$ such that $y([0, t]) \in M^x_{a,b}(t)$ for all $t \in [0, T]$.

Let $g_L = g_L(x)$ be the maximum of the Lipschitz constants of all g_j on $M^x_{a,b}(T)$. Let us choose a constant $K = K(x) \geq g_L$. By conditional positivity, $g_j(y) \geq -Ky_j$ in $M^x_{a,b}(T)$, because

$$\|g_j(y) - g_j(y_1, \cdots, y_{j-1}, 0, y_{j+1}, \cdots)\| \leq Ky_j \text{ and } g_j(y_1, \cdots, y_{j-1}, 0, y_{j+1}, \cdots) \geq 0.$$

Hence by rewriting equation (4.2) equivalently as

$$\dot{y} = (g(y) + Ky) - Ky \iff \{\dot{y}_j = (g_j(y) + Ky_j) - Ky_j, \quad j = 1, \cdots, n\}, \quad (4.6)$$

we ensure that the "nonlinear part" $g_j(y) + Ky_j$ of the r.h.s. is always nonnegative.

We modify the usual approximation scheme (see the remark above) for ODEs by defining the map Φ_x from $C_{a,b}(T)$ to itself in the following way: for $y \in C([0, T], \mathbf{R}_n^+)$, let $\Phi_x(y)$ be the solution of the equation

$$\frac{d}{dt}[\Phi_x(y)](t) = g(y(t)) + Ky(t) - K[\Phi_x(y.)](t),$$

with the initial data $[\Phi_x(y)](0) = x$. It is a linear equation with a unique explicit solution, which can be taken as an alternative definition of Φ_x:

$$[\Phi_x(y)](t) = e^{-Kt}x + \int_0^t e^{-K(t-s)}[g(y(s)) + Ky(s)]\,ds.$$

Clearly, Φ_x preserves positivity, and fixed points of Φ are positive solutions to (4.6) with the initial data x.

Let us check that Φ takes $C_{a,b}(T)$ to itself. If $y \in C_{a,b}(T)$, then

$$(L, [\Phi_x(y)](t))) = e^{-Kt}(L, x) + \int_0^t e^{-K(t-s)}[(a + K)(L, y(s)) + b]\,ds$$

$$\leq e^{-Kt}(L, x) + \int_0^t e^{-K(t-s)}[(a + K)\left(e^{as}\left((L, x) + \frac{b}{a}\right) - \frac{b}{a}\right) + b]\,ds$$

$$= (L, x)e^{-Kt} + e^{-Kt}(e^{(K+a)t} - 1)((L, x) + b/a) - (b/a)e^{-Kt}(e^{Kt} - 1)$$

$$= (L, x)e^{at} + \frac{b}{a}(e^{at} - 1).$$

Notice that due to this bound, the iterations of Φ remain in $C_{a,b}(T)$, and hence it is justified to use the Lipschitz constant K for g.

Next, the map $\Phi(t)$ is a contraction for small t, because

$$\|[\Phi_{x_1}(y^1)](t) - [\Phi_{x_2}(y^2)](t)\| \leq (K + g_L)\int_0^t \|y_s^1 - y_s^2\|\,ds + \|x_1 - x_2\|,$$

and the proof of the existence and uniqueness of a fixed point is completed by the usual application of the Banach fixed-point principle.

Finally,

$$\|X(t, x) - X(t, y)\| \leq \int_0^t \varkappa(\lambda(t))\|X(s, x) - X(s, y)\|\,ds + \|x - y\|,$$

and (4.5) follows by the usual Grönwall's lemma. $\square$

Once the solution $X(t, x)$ is constructed, the linear operators T^t,

$$T^t f(x) = f(X(t, x)), \quad t \geq 0, \tag{4.7}$$

become well-defined contractions in $C(\mathcal{M}^+(L))$ forming a semigroup. In case $a = b = 0$, the operators U^t form a semigroup of contractions also in $C(\mathcal{M}^+_{\leq \lambda}(L))$ for all λ.

Theorem 4.1.2. *Under the assumptions of Theorem 4.1.1, assume additionally that g is twice continuously differentiable on $\mathcal{M}^+(L)$, so that for all λ,*

$$\|g^{(2)}(\mathcal{M}^+_{\leq \lambda}(L))\| \leq D_2(\lambda) \tag{4.8}$$

with some continuous function D_2. Then the solutions to $X(t, x)$ from Theorem 4.1.1 are twice continuously differentiable with respect to initial data and

$$\|X(t, \cdot)^{(1)}(\mathcal{M}^+_{\leq \lambda}(L))\| \leq e^{t \varkappa(\lambda(t))}, \tag{4.9}$$

$$\|X(t, \cdot)^{(2)}(\mathcal{M}^+_{\leq \lambda}(L))\| = \sup_{j,i} \sup_{x \in \mathcal{M}^+_{\leq \lambda}(L)} \|\frac{\partial^2 X(t, x)}{\partial x_i \partial x_j}\| \leq t D_2(\lambda(t)) e^{3t \varkappa(\lambda(t))}. \tag{4.10}$$

Moreover,

$$\|(T^t f)^{(1)}(\mathcal{M}^+_{\leq \lambda}(L))\| = \sup_{j} \sup_{x \in \mathcal{M}^+_{\leq \lambda}(L)} \left| \frac{\partial}{\partial x_j} f(X(t, x)) \right| \leq \|f^{(1)}(\mathcal{M}^+_{\leq \lambda}(L))\| e^{t \varkappa(\lambda(t))},$$

$$\tag{4.11}$$

$$\|(T^t f)^{(2)}(\mathcal{M}^+_{\leq \lambda}(L))\| = \sup_{j,i} \sup_{x \in \mathcal{M}^+_{\leq \lambda}(L)} \left| \frac{\partial^2}{\partial x_j \partial x_i} f(X(t, x)) \right|$$

$$\leq \|f^{(2)}(\mathcal{M}^+_{\leq \lambda(t)}(L))\| e^{2t \varkappa(\lambda(t))} + t \|f^{(1)}(\mathcal{M}^+_{\leq \lambda(t)}(L))\| D_2(\lambda(t)) e^{3t \varkappa(\lambda(t))}. \tag{4.12}$$

Proof. Differentiating the equation $\dot{x} = g(x)$ with respect to the initial conditions yields

$$\frac{d}{dt} \frac{\partial X(t, x)}{\partial x_j} = \sum_k \frac{\partial g}{\partial x_k}(X(t, x)) \frac{\partial X_k(t, x)}{\partial x_j}, \tag{4.13}$$

implying (4.9). Differentiating the equation $\dot{x} = f(x)$ twice yields

$$\frac{d}{dt} \frac{\partial^2 X(t, x)}{\partial x_j \partial x_i} = \sum_k \frac{\partial g}{\partial x_k}(X(t, x)) \frac{\partial^2 X_k(t, x)}{\partial x_j \partial x_i} + \sum_{k,l} \frac{\partial^2 g}{\partial x_k \partial x_l}(X(t, x)) \frac{\partial X_k(t, x)}{\partial x_j} \frac{\partial X_l(t, x)}{\partial x_i}.$$

$$\tag{4.14}$$

Solving this linear equation with the initial condition $\partial^2 X(0, x)/\partial x_j \partial x_i = 0$ yields (4.10).

Differentiating (4.7) yields

$$\frac{\partial}{\partial x_j}(T^t f)(x) = \sum_k \frac{\partial f}{\partial x_k}(X(t,x))\frac{\partial X_k(t,x)}{\partial x_j}, \tag{4.15}$$

implying (4.11). Differentiating a second time yields

$$\frac{\partial^2}{\partial x_j \partial x_i}(T^t f)(x) = \sum_k \frac{\partial f}{\partial x_k}(X(t,x))\frac{\partial^2 X_k(t,x)}{\partial x_j \partial x_i} + \sum_{k,l} \frac{\partial^2 f}{\partial x_k \partial x_l}(X(t,x))\frac{\partial X_k(t,x)}{\partial x_j}\frac{\partial X_l(t,x)}{\partial x_j}, \tag{4.16}$$

implying (4.12). □

Remark 19. *The proof above is not complete. All estimates are proved based on the assumption that the required derivatives exist. However, the corresponding justification arguments are standard in the theory of ODEs and thus are not reproduced here; see, e.g., [142].*

4.2   Mean-Field Interacting Systems in l^1

In this chapter we extend the main results of Chapters 2 and 3 in two directions, namely, by working with a countable (rather than finite or compact) state space and unbounded rates, and with more general interactions allowing in particular for a change in the number of particles.

Therefore, we take the set of natural numbers $\{1, 2, \cdots\}$ as the state space of each small player, the set of finite Borel measures on it being the Banach space l^1 of summable real sequences $x = (x_1, x_2, \cdots)$.

Thus the state space of the total multitude of small players will be formed by the set $\mathbf{Z}_+^{fin}$ of sequences of integers $n = (n_1, n_2, \cdots)$ with only a finite number of nonzero elements, with n_k denoting the number of players in the state k, the total number of small players being $N = \sum_k n_k$. Since we are going to extend the analysis to processes not preserving the number of particles, we shall work now with a more general scaling of the states, namely with the sequences

$$x = (x_1, x_2, \cdots) = hn = h(n_1, n_2, \cdots) \in h\mathbf{Z}_+^{fin}$$

with the parameter $h = 1/N_0$, the inverse of the total number of players $\sum_k n_k$ at the initial moment of observation. The necessity to distinguish the initial moment is crucial here, since this number changes over time. Working with the scaling related to the current number of particles N may lead, of course, to different evolutions.

The general processes of birth, death, mutations, and binary interactions that can occur under the effort b of the principal are Markov chains on $h\mathbf{Z}_+^{fin}$ specified by the generators of the following type:

$$L_{b,h}F(x) = \frac{1}{h}\sum_j \beta_j(x,b)[F(x+he_j)-F(x)]$$

$$+\frac{1}{h}\sum_j \alpha_j(x,b)[F(x-he_j)-F(x)]$$

$$+\frac{1}{h}\sum_{i,j}\alpha_{ij}^1(x,b)[F(x-he_i+he_j)-F(x)]$$

$$+\frac{1}{h}\sum_{i,(j_1,j_2)}\alpha_{i(j_1 j_2)}^1(x,b)[F(x-he_i+he_{j_1}+he_{j_2})-F(x)]$$

$$+\frac{1}{h}\sum_{(i_1 i_2),j}\alpha_{(i_1 i_2)j}^2(x,b)[F(x-he_{i_1}-he_{i_2}+he_j)-F(x)]$$

$$+\frac{1}{h}\sum_{(i_1,i_2)}\sum_{(j_1,j_2)}\alpha_{(i_1 i_2)(j_1 j_2)}^2(x,b)[F(x-he_{i_1}-he_{i_2}+he_{j_1}+he_{j_2})-F(x)],$$

$$(4.17)$$

where parentheses (i,j) denote the pairs of states. Here the terms with β_j and α_j describe the spontaneous injection (birth) and death of agents, the terms with α_{ij}^1 describe the mutation or migration of single agents, the terms with $\alpha_{i(lj)}^1$ describe the fragmentation or splitting, the terms with α^2 describe the binary interactions, which can result in either merging of two agents, producing an agent with another strategy, or their simultaneous migration to any other pair of strategies. All terms include possible mean-field interactions. For example, our model (2.31) was an example of migration.

Let L be a strictly positive nondecreasing function on $\mathbf{N}$. We shall refer to such functions as Lyapunov functions. We say that the generator $L_{b,h}$ with $\beta_j = 0$ and the corresponding process do not increase L if for every allowed transition, the total value of L cannot increase; that is, if $\alpha_{ij}^1 \neq 0$, then $L(j) \leq L(i)$, if $\alpha_{i(j_1,j_2)}^1 \neq 0$, then $L(j_1)+L(j_2) \leq L(i)$, if $\alpha_{(i_1 i_2)j}^2 \neq 0$, then $L(j) \leq L(i_1)+L(i_2)$, if $\alpha_{(i_1 i_2)(j_1 j_2)}^2 \neq 0$, then $L(j_1)+L(j_2) \leq L(i_1)+L(i_2)$. If this is the case, the chains generated by $L_{b,h}$ always remain in $\mathcal{M}_{\leq\lambda}^+(L)$ if they were started there. Hence if $L_j \to \infty$ as $j \to \infty$, then the sets $h\mathbf{Z}_+^{fin} \cap \mathcal{M}_{\leq\lambda}^+(L)$ are finite for all h and λ, and $L_{b,h}$ generates well-defined Markov chains $X_{b,h}(t,x)$ in each of these sets.

A generator $L_{b,h}$ is called L-subcritical if $L_{b,h}(L) \leq 0$. Of course, if $L_{b,h}$ does not increase L, then it is L-subcritical. Although the condition to not increase L seems to be restrictive, many concrete models satisfy it, for instance, the celebrated merging–splitting (Smoluchowski) process considered below. On the other hand, models with spontaneous injections may increase L, so that one is restricted to working with the weaker property of subcriticality. We shall first analyze the case of chains that do not increase L, and then will consider subcritical evolutions and evolutions allowing for a mild growth of L.

By Taylor-expanding F in (4.17), one sees that if F is sufficiently smooth, then the sequence $L_{b,h}F$ converges to

$$\Lambda_b F(x) = \sum_j (\beta_j(x,b) - \alpha_j(x,b))\frac{\partial F}{\partial x_i} + \sum_{i,j} \alpha_{ij}^1(x,b)[\frac{\partial F}{\partial x_j} - \frac{\partial F}{\partial x_i}]$$

$$+ \sum_{i,(j_1,j_2)} \alpha_{i(j_1 j_2)}^1(x,b)[\frac{\partial F}{\partial x_{j_1}} + \frac{\partial F}{\partial x_{j_2}} - \frac{\partial F}{\partial x_i}] + \sum_{(i_1,i_2)} \sum_j \alpha_{(i_1 i_2)j}^2(x,b)[\frac{\partial F}{\partial x_j} - \frac{\partial F}{\partial x_{i_1}} - \frac{\partial F}{\partial x_{i_2}}]$$

$$+ \sum_{(i_1,i_2)} \sum_{(j_1,j_2)} \alpha_{(i_1 i_2)(j_1 j_2)}^2(x,b)[\frac{\partial F}{\partial x_{j_1}} + \frac{\partial F}{\partial x_{j_2}} - \frac{\partial F}{\partial x_{i_1}} - \frac{\partial F}{\partial x_{i_2}}]. \tag{4.18}$$

Moreover,

$$\|(L_{b,h} - \Lambda_b)F\|_{C(\mathcal{M}_{\leq\lambda}^+(L))} \leq 8h\varkappa(\lambda)\|F\|_{C^2(\mathcal{M}_{\leq\lambda}^+(L))}, \tag{4.19}$$

with $\varkappa(L,\lambda)$ the $\sup_b$ of the norms

$$\left\|\sum_i (\alpha_i + \beta_i) + \sum_{i,j} \alpha_{ij}^1 + \sum_{i,(j_1,j_2)} \alpha_{i(j_1 j_2)}^1 + \sum_{(i_1,i_2),j} \alpha_{(i_1 i_2)j}^2 + \sum_{(i_1,i_2),(j_1,j_2)} \alpha_{(i_1 i_2)(j_1 j_2)}^2\right\|_{C(\mathcal{M}_{\leq\lambda}^+(L))}. \tag{4.20}$$

By regrouping the terms of Λ_b, it can be rewritten in the form of the general first-order operator

$$\Lambda_b F(x) = \sum_j g_j(x,b)\frac{\partial F}{\partial x_j}, \tag{4.21}$$

where $g = (g_i)$ with

$$g_i = \beta_i - \alpha_i + \sum_k (\alpha_{ki}^1 - \alpha_{ik}^1) + \sum_k [\alpha_{k(ii)}^1 + \sum_{j\neq i}(\alpha_{k(ij)}^1 + \alpha_{k(ji)}^1] - \sum_{(j_1,j_2)} \alpha_{i(j_1 j_2)}^1$$

$$+ \sum_{(j_1,j_2)} \alpha_{(j_1 j_2)i}^2 - \sum_k [\alpha_{(ii)k}^2 + \sum_{j\neq i}(\alpha_{(ij)k}^2 + \alpha_{(ji)k}^2)]$$

$$+ \sum_{(j_1,j_2)} [\alpha_{(j_1 j_2)(ii)}^2 + \sum_{j\neq i}(\alpha_{(j_1 j_2)(ji)}^2 + \alpha_{(j_1 j_2)(ij)}^2)]$$

$$- \sum_{(j_1,j_2)} [\alpha_{(ii)(j_1 j_2)}^2 + \sum_{j\neq i}(\alpha_{(ij)(j_1 j_2)}^2 + \alpha_{(ji)(j_1 j_2)}^2)].$$

Its characteristics solving the ODE $\dot{x} = g(x)$ can be expected to describe the limiting behavior of the Markov chains $X_{b,h}(x,t)$ for $h \to 0$.

As in the finite-dimensional case, we can build the semigroups of transition operators $U_h^t F(x) = \mathbf{E}F(X_h(t,x))$, $U^t F(x) = \mathbf{E}F(X(t,x))$, and we are going to search for conditions ensuring that $U_h^t \to U^t$ as $h \to 0$.

Let us stress again that the present models are much more general than the models of Chapter 2. The present g cannot always be written in the form $g = x Q(x)$ with a Q-matrix Q describing the evolution of one player, because the possibility of birth and death makes the tracking of a particular player throughout the game impossible.

4.3 Convergence Results for Evolutions in l^1

As in Chapter 2, let us begin with the case of smooth coefficients.

Theorem 4.3.1. *Assume that the operators $L_{b,h}$ are L nonincreasing for a (positive nondecreasing) Lyapunov function L on $\mathbf{Z}$, and the function $g : \mathcal{M}^+_{\leq\lambda}(L) \to l^1$ belongs to $C^2(\mathcal{M}^+_{\leq\lambda}(L))$. Then the Markov chains $X_h(t, x(h))$ with $x(h) \in \mathcal{M}^+_{\leq\lambda}(L)$ such that $x(h) \to x$ as $h \to 0$ converge in distribution to the deterministic evolution $X(t, x)$, and moreover,*

$$\|U^t_h F - U^t F\| = \sup_x |\mathbf{E} F(X_h(t, x)) - F(X(t, x))|$$

$$\leq 8 t h \varkappa(L, \lambda)(\|F^{(2)}\| + t \|F^{(1)}\| \|g^{(2)}\|) e^{3t\|g\|_{Lip}}, \tag{4.22}$$

where all norms are understood as the $\sup_b$ of the norms of functions on $\mathcal{M}^+_{\leq\lambda}(L) \subset l^1$.

Proof. The proof is the same as the proof of Theorem 2.4.1.

All function norms below are the norms of functions defined on $\mathcal{M}^+_{\leq\lambda}(L)$. By (4.11),

$$\|(U^t F)^{(2)}\| \leq (\|F^{(2)}\| + t \|F^{(1)}\| \|g^{(2)}\|) e^{3t\|g\|_{Lip}}, \tag{4.23}$$

because $\lambda(t) = \lambda$ (this is precisely the simplifying assumption that all transitions do not increase L). Hence by (4.19),

$$\|(L_{b,h} - \Lambda_b) U^t F\| \leq 8 h \varkappa(L, \lambda)(\|F^{(2)}\| + t \|F^{(1)}\| \|g^{(2)}\|) e^{3t\|g\|_{Lip}},$$

implying (4.22). $\qquad\square$

Let us turn to the case of Lipschitz continuous coefficients.

Theorem 4.3.2. *Assume that the operators $L_{b,h}$ are L nonincreasing for a Lyapunov function L on $\mathbf{Z}$ such that $L(j)/j^\gamma \to 1$ with $\gamma > 0$ as $j \to \infty$. Let $g : \mathcal{M}^+_{\leq\lambda}(L) \to l^1$ belong to $(\mathcal{M}^+_{\leq\lambda}(L))_{bLip}$. Then the Markov chains $X_h(t, x(h))$ with $x(h) \in \mathcal{M}^+_{\leq\lambda}(L)$ such that $x(h) \to x$ as $h \to 0$ converge in distribution to the deterministic evolution $X(t, x)$, and moreover,*

$$|\mathbf{E} F(X_h(t, x)) - F(X(t, x))| \leq C \|F\|_{Lip}(t \|g\|_{bLip} + \varkappa(L, \lambda))(ht)^{\gamma/(1+2\gamma)} \tag{4.24}$$

with a constant C. In particular, if $L(j) = j$, as is often the case in applications,

$$|\mathbf{E}F(X_h(t, x)) - F(X(t, x))| \leq C\|F\|_{Lip}(t\|g\|_{bLip} + \varkappa(L, \lambda))(ht)^{1/3}. \quad (4.25)$$

Proof. The proof is similar to the proof of Theorem 2.5.1. The new difficulty arises from the fact that the number of pure states (which is the maximal j such that the sequence $(\delta_i^j) \in l^1$ belongs to $\mathcal{M}_{\leq\lambda}^+(L) \cap h\mathbf{Z}_+^\infty$) increases with a decrease in h, while in Theorem 2.5.1, the number of pure states d was fixed. Thus the smoothing will be combined with finite-dimensional approximations.

Let P_j denote the projection on the first j coordinates in $\mathbf{R}^\infty$, that is, $[P_j(x)]_k = x_k$ for $k \leq j$ and $[P_j(x)]_k = 0$ otherwise. For $x \in \mathcal{M}_{\leq\lambda}(L)$,

$$\|P_j(x) - x\|_{l^1} \leq \frac{\lambda}{L(j)}.$$

For functions on $x \in \mathcal{M}_{\leq\lambda}(L)$, we shall use the smooth approximation $\Phi_{\delta,j}(F) = \Phi_\delta(F \circ P_j)$ with Φ_δ as in Theorem 2.5.1. More precisely,

$$\Phi_{\delta,j}[V](x) = \int_{R^j} \frac{1}{\delta^j} \prod_{k=1}^j \chi\left(\frac{y_k}{\delta}\right) V(x - y) \prod_{k=1}^j dy_k = \int_{\mathbf{R}^j} \frac{1}{\delta^j} \prod_{k=1}^j \chi\left(\frac{x_k - y_k}{\delta}\right) V(y) \prod_{k=1}^j dy_k,$$

where χ is a mollifier, a nonnegative infinitely smooth even function on $\mathbf{R}$ with compact support and $\int \chi(w)\, dw = 1$. Thus

$$\Phi_{\delta,j}[V](x) = \int_{R^j} \frac{1}{\delta^j} \prod_{k=1}^j \chi\left(\frac{y_k}{\delta}\right) V(P_j(x - y)) \prod_{k=1}^j dy_k.$$

The estimates (2.60) and (2.62) of Theorem 2.5.1 remain the same:

$$\|\Phi_{\delta,j}[V]\|_{C^1} = |\Phi_\delta[V]\|_{bLip} \leq \|V\|_{bLip}$$

and

$$\|\Phi_{\delta,j}[V]^{(2)}\| \leq \|V\|_{bLip} \frac{1}{\delta} \int |\chi'(w)|\, dw \quad (4.26)$$

for all δ, and (2.61) generalizes to

$$\|\Phi_{\delta,j}[V] - V\| \leq C\|V\|_{Lip}(j\delta + \lambda/L(j)). \quad (4.27)$$

Consequently, arguing as in Theorem 2.5.1, we obtain (using similar notation)

$$\|U_{h,\delta,j}^t f - U_h^t f\| \leq Ct\|Q\|_{bLip}\|f\|_{bLip}(\delta j + \lambda/L(j)) \exp\{t\|g\|_{Lip}\},$$

$$\|U_{\delta,j}^t f - U^t f\| \leq Ct\|Q\|_{bLip}\|f\|_{bLip}(\delta j + \lambda/L(j)) \exp\{t\|g\|_{Lip}\},$$

$$\|U_{h,\delta,j}^t f - U_{\delta,j}^t f\| \le 8th\varkappa(L,\lambda)(\|F^{(2)}\| + t\|F\|_{Lip}\|g^{(2)}\|)e^{3t\|g\|_{Lip}}.$$

And therefore,

$$\|U_h^t F - U^t F\| \le 8th\varkappa\|F^{(2)}\|e^{3t\|g\|_{Lip}} +$$

$$Ct\|F\|_{Lip}\|g\|_{bLip}\left[\frac{ht}{\delta}\varkappa(L,\lambda) + j\delta + \frac{\lambda}{L(j)}\right]e^{3t\|g\|_{Lip}}.$$

To get rid of $F^{(2)}$, we also approximate F by the function $\Phi_{\delta,j}[F]$, so that

$$\|U_h^t F - U^t F\| \le C\|F\|_{Lip}(t\|g\|_{bLip} + \varkappa(L,\lambda))\left[\frac{ht}{\delta}\varkappa(L,\lambda) + j\delta + \frac{\lambda}{L(j)}\right]e^{3t\|g\|_{Lip}}.$$

Choosing

$$j = (ht)^{-1/(1+2\gamma)}, \delta = (ht)^{(1+\gamma)/(1+2\gamma)}$$

makes all three terms in the square brackets of the same order, yielding (4.24). $\square$

Let us now write down the analogue of the multistep optimization result of Theorem 3.1.1. Namely, assume that the principal is updating its strategy in discrete times $\{k\tau\}, k = 0, 1, \cdots, n-1$, with some fixed $\tau > 0, n \in \mathbf{N}$, aiming at finding a strategy π maximizing the reward (3.2), but now with $x_0 = x(h) \in h\mathbf{Z}^{fin} \cap \mathcal{M}_{\le\lambda}^+(L)$ and the transitions $x_k = X_h(\tau, x_{k-1}, b_{k-1})$ defined by the Markov chain generated by (4.17). The Shapley operators SN and S are defined analogously to their definition in Section 3.1, so that their iterations define the optimal payoffs for the random evolution with fixed h and for the limiting deterministic evolution.

Theorem 4.3.3. *Under the assumption of Theorem 4.3.2, let $x_0 = x(h) \in h\mathbf{Z}^{fin} \cap \mathcal{M}_{\le la}^+(L)$ and $x(h) \to x$ in l^1. Then for all $\tau \in (0, 1], n \in N$, and $t = \tau n$,*

$$\|S^n[N]V(x(h)) - S^n V(x)\|$$
$$\le C(t, \|g\|_{bLip}, \|V\|_{bLip})(h^{\gamma/(1+2\gamma)}\tau^{-(1+\gamma)/(1+2\gamma)} + |x(h) - x|), \quad (4.28)$$

with a constant C depending on $(t, \|g\|_{bLip}, \|V\|_{bLip})$. The r.h.s. tends to zero if $h \to 0$ faster than $\tau^{1+1/\gamma}$.

All other results of Chapter 3 extend automatically (with appropriate modifications) to the case of Markov chains generated by (4.17).

4.4 Evolutionary Coalition-Building Under Pressure

As a direct application of Theorems 4.3.2 and 4.3.3, let us discuss the model of evolutionary coalition-building. Namely, so far, we have talked about small players that occasionally and randomly exchange information in small groups (mostly in randomly formed pairs) resulting in copying the most successful strategy by the members of the group. Another natural reaction of the society of small players to the pressure exerted by the principal can be executed by forming stable groups that can confront this pressure in a more effective manner (but possibly imposing certain obligatory regulations for the members of the group). Analysis of such a possibility leads one naturally to models of mean-field-enhanced coagulation processes under external pressure. Coagulation–fragmentation processes are well studied in statistical physics; see, e.g., [184]. In particular, general mass-exchange processes, which in our social environment become general coalition-forming processes preserving the total number of participants, were analyzed in [129] and [131] with their law of large number limits for discrete and general state spaces. Here we add to this analysis a strategic framework for a major player, fitting the model to the more general framework of the previous section. Instead of coagulation and fragmentation of particles, we shall use here the terms merging and splitting (or breakage) of the coalitions of agents.

For simplicity, we ignore here any other behavioral distinctions (assuming no strategic space for an individual player), concentrating only on the process of forming coalitions. Thus the state space of the total multitude of small players will be formed by the set $\mathbf{Z}_+^{fin}$ of sequences of integers $n = (n_1, n_2, \cdots)$ with only finite number of nonvanishing elements, with n_k denoting the number of coalitions of size k, the total number of small players being $N = \sum_k k n_k$ and the total number of coalitions (a single player is considered to represent a coalition of size 1) being $\sum_k n_k$. Also, for simplicity, we restrict attention to binary merging and breakage only; the extension to arbitrary regrouping processes from [129] (preserving the number of players) is more or less straightforward.

As previously, we will look for the evolution of appropriately scaled states, namely the sequences

$$x = (x_1, x_2, \cdots) = hn = h(n_1, n_2, \cdots) \in h\mathbf{Z}_+^{fin}$$

with a certain parameter $h > 0$, which can be taken, for instance, as the inverse of the total number $\sum_k n_k$ of coalitions at the initial moment of observation.

If any randomly chosen pair of coalitions of sizes j and k can merge with the rate $C_{kj}(x, b)$, which may depend on the whole composition x and the control parameter b of the major player, and any randomly chosen coalition of size j can split (break, fragment) into two groups of sizes $k < j$ and $j - k$ with rate $F_{jk}(x, b)$, then the limiting deterministic evolution of the state is known to be described by the system of the so-called Smoluchowski equations

$$\dot{x}_k = g_k(x) = \sum_{j<k} C_{j,k-j}(x, b) x_j x_{k-j} - 2 \sum_j C_{kj}(x, b) x_j x_k$$
$$+ 2 \sum_{j>k} F_{jk}(x, b) x_j - \sum_{j<k} F_{kj}(x, b) x_k. \tag{4.29}$$

In addition to the well-known setting with constant C_{jk} and F_{jk} (see, e.,g., [24]), we have added here the mean-field dependence of these coefficients (dependence on x) and the dependence on the control parameter b.

As one easily checks, equation (4.29) can be written in the equivalent weak form

$$\frac{d}{dt} \sum_j \phi_j x_j = \sum_{j,k} (\phi_{j+k} - \phi_j - \phi_k) C_{jk}(x,b) x_j x_k + \sum_j \sum_{k<j} (\phi_{j-k} + \phi_k - \phi_j) F_{jk}(x,b) x_j,$$

(4.30)

which should hold for a suitable class of test functions ϕ. For instance, under the assumption of bounded coefficients (see (4.35) below), the class of test functions is the class of all functions from $l^\infty = \{\phi : \sup_j |\phi_j| < \infty\}$. This implies, in particular, that the corresponding semigroups (4.7) on the space of continuous functions, that is, $U^t G(x) = G(X(t,x))$, have the generator

$$\Lambda_b G(x) = \sum_k g_k(x) \frac{\partial G}{\partial x_k}(x) = \sum_{j,k} \left(\frac{\partial G}{\partial x_{k+j}} - \frac{\partial G}{\partial x_j} - \frac{\partial G}{\partial x_k} \right) C_{jk}(x,b) x_j x_k$$

$$+ \sum_j \sum_{k<j} \left(\frac{\partial G}{\partial x_{j-k}} - \frac{\partial G}{\partial x_j} + \frac{\partial G}{\partial x_k} \right) F_{jk}(x,b) x_j$$

(4.31)

of type (4.18) with

$$\alpha^1_{i(j_1 j_2)} = F_{ij_1}(x) x_i \quad \text{for} \quad j_1 + j_2 = i; \quad \alpha^2_{(i_1 i_2)j} = C_{i_1 i_2}(x) x_{i_1} x_{i_2} \quad \text{for} \quad j = i_1 + i_2$$

and all other coefficients vanishing. Thus the Smoluchowski equations are particular representatives of the characteristics of the first-order PDE (4.18) or (4.21).

Let $R_j(x,b)$ be the payoff for the member of a coalition of size j. In our strategic setting, the rates $C_{jk}(x,b)$ and $F_{jk}(x,b)$ should depend on the differences in these rewards before and after merging or splitting. For instance, the simplest choices can be

$$C_{kj}(x,b) = a_{j+k,k}(R_{k+j} - R_k)^+ + a_{j+k,j}(R_{k+j} - R_j)^+,$$

(4.32)

with some constants $a_{lk} \geq 0$ reflecting the assumption that merging may occur whenever it is beneficial for all members concerned but weighted according to the size of the coalitions involved. Similarly,

$$F_{kj}(x,b) = \tilde{a}_{kj}(R_j - R_k)^+ + \tilde{a}_{k,k-j}(R_{k-j} - R_k)^+.$$

(4.33)

A Markov approximation to the dynamics (4.29) is constructed in the standard way, which is analogous to the constructions of approximating Markov chains described in the previous section. (for coagulation–fragmentation processes, this Markov approximation is often referred to as the Markus–Lushnikov process; see, e.g., [184]), namely, by attaching exponential clocks to every pair of coalitions that can merge with rates C_{kj} and to every coalition that can split with rates F_{kj}. This leads to the Markov chain $X_h(t, x, b)$ on $h\mathbf{Z}_+^{fin}$ describing the process of *coalition formation* with the generator

$$L_{b,h}^{coal} G(x) = \frac{1}{h} \sum_{i,j} C_{ij}(x, b) x_i x_j [G(x - he_i - he_j + he_{i+j}) - G(x)]$$

$$+ \frac{1}{h} \sum_i \sum_{j<i} F_{ij}(x, b) x_i [G(x - he_i + he_j + he_{i-j}) - G(x)], \qquad (4.34)$$

of type (4.17), with the same identification of coefficients as above.

We shall propose here only the simplest result in this direction, assuming that the intensities of individual transitions are uniformly bounded and uniformly Lipschitz, that is,

$$C = \sup_{j,k} C_{jk}(x, b) < \infty, \quad F = \sup_j \sum_{k<j} F_{kj}(x, b) < \infty, \qquad (4.35)$$

$$C(1) = \sup_{b,j,k} \|C_{jk}(., b)\|_{C_{bLip}(\mathcal{M}_{\leq\lambda}^+(L))} < \infty,$$

$$F(1) = \sup_{b,j} \sum_{k<j} \|F_{kj}(., b)\|_{C_{bLip}(\mathcal{M}_{\leq\lambda}^+(L))} < \infty, \qquad (4.36)$$

with the Lyapunov function $L(j) = j$. Notice, however, that the overall intensities are still unbounded (quadratic), so that we are still far beyond the assumptions of Chapter 2.

For the function $L(j) = j$, the Markov chains $X_h(t, x, b)$ do not increase L. Moreover, (4.35) implies

$$\|g\| \leq 3C\lambda^2 + 3F\lambda,$$

$$\|g\|_{bLip} \leq 6C\lambda + 3F + 3(C(1)\lambda + F(1))\lambda, \qquad (4.37)$$

where we use the convention of the previous section that all functional norms are understood as $\sup_b$ of the norms of functions on $\mathcal{M}_{\leq\lambda}^+(L)$ with some fixed λ. Hence the following result holds.

Theorem 4.4.1. *For a model of strategically enhanced coalition-building subject to (4.35) and (4.36), the conditions of Theorems 4.3.2 and 4.3.3 are satisfied for operators (4.31) and (4.34), which represent particular cases of operators (4.18) and (4.17). Hence the corresponding Markov chains $X_h(t, x, b)$ converge to the deterministic limit governed by the equation $\dot{x} = f(x, b)$, and the corresponding multistep Markov decision problem converges to its deterministic limit.*

4.5   Strategically Enhanced Preferential Attachment on Evolutionary Background

A natural and useful extension of the theory presented above can be obtained by the inclusion in our pressure–resistance evolutionary-type game the well-known model of linear growth with preferential attachment (Yule, Simon, and others; see [216] for a review), turning the latter into a strategically enhanced preferential attachment model that includes evolutionary-type interactions between agents and a major player having tools to control (interfere with) this interaction.

We shall work with the general framework of Theorem 4.3.2, having in mind that the basic examples of the approximating Markov chains $X_h(t, x(h))$ can arise from the merging and splitting coalition model of the previous section (with generator (4.34)) or from the mean-field interacting particle systems of Chapters 1 and 2 (for instance, 1.10), where the number of possible states becomes infinite, which can be looked at as describing the evolution of coalitions resulting from individual migrations from one coalition to another. If we denote by j the size of the coalitions (rather than the type of an agent, as in (1.10)), then the generator of this Markov chain becomes an infinite-dimensional version of (2.33):

$$L_{b,h}^{migr} G(x) = \frac{1}{h} \sum_{i,j} \varkappa x_i x_j [R_j(x, b) - R_i(x, b)]^+ [G\left(x - he_i + he_j\right) - G(x)],$$

$$(4.38)$$

where $R_j(x, b)$ is the payoff to a member of a coalition of size $j = 1, 2, \cdots$. The Markov chain with generator (4.38) describes the process whereby agents can move from one coalition to another, choosing the size of the coalition that is more beneficial under the control b of the principal.

The most-studied form of preferential attachment evolves by the discrete-time injections of agents (see [27, 69, 216], and references therein). Along these lines, we can assume that with time intervals τ, a new agent enters the system in such a way that with some probability $p(x, b)$ (which, in contrast to the standard model, can now depend on the distribution x and the control parameter b of the principal), it does not enter any of the existing coalitions (thus forming a new coalition of size 1), and with probability $1 - p(x, b)$, it joins one of the coalitions, the probability of joining a coalition being proportional to its size (this reflects the notion of preferential attachment coined in [27]).

A continuous-time version of these evolutions can be modeled by a Markov process, where the injection occurs with some intensity $\alpha(x, b)$ (which can be influenced by the principal, subject to certain costs). In other words, it can be included by adding to the generator (4.38) or (4.34) an additional term of type

$$L_{b,h}^{att} G(x) = \frac{p(x, b)\alpha(x, b)}{h} [G(x + he_1) - G(x)]$$

$$+\frac{(1-p(x,b))\alpha(x,b)}{h}\sum_{k=1}^{\infty}kx_k[G(x-he_k+he_{k+1})-G(x)],$$

or, setting $a(x,b)=p(x,b)\alpha(x,b)$, $P(x,b)=(1-p(x,b))\alpha(x,b)$,

$$L_{b,h}^{att}G(x)=\frac{a(x,b)}{h}[G(x+he_1)-G(x)]+\frac{P(x,b)}{h}\sum_{k=1}^{\infty}kx_k[G(x-he_k+he_{k+1})-G(x)].$$

$$(4.39)$$

Division by $h=1/N_0$ in this equation makes the rate of growth of the number of agents comparable to the total number of agents. If it were of smaller scale, it would not influence significantly the limiting LLN evolution.

The limiting generator obtained form (4.39) as $h\to 0$ is the first-order operator

$$\Lambda_{b,h}^{att}G(x)=a(x,b)\frac{\partial G}{\partial x_1}+P(x,b)\sum_{k=1}^{\infty}kx_k\left[\frac{\partial G}{\partial x_{k+1}}-\frac{\partial G}{\partial x_k}\right],\qquad(4.40)$$

with the characteristics of the kinetic equations

$$\dot{x}_j=\delta_1^j a(x,b)+P(x,b)[(j-1)x_{j-1}-jx_j].\qquad(4.41)$$

Combining this with the processes given by (4.38) or (4.34) yields the model of *coalition-building with the influx of new agents* or the *strategically enhanced preferential attachment model on an evolutionary background*. To shorten the formulas, let us ignore the contribution of individual migration (4.38) and concentrate on the combination of (4.34) and (4.40), that is, on the chain $X_h(t,x(h))$ generated by the operator

$$L_{b,h}=L_{b,h}^{att}+L_{b,h}^{coal},\qquad(4.42)$$

with the limiting generator $\Lambda_{b,h}=\Lambda_{b,h}^{att}+\Lambda_{b,h}^{coal}$. The corresponding characteristics represent the controlled (via discrete- or continuous-time choice of parameter b by the principal) infinite-dimensional ODEs combining (4.41) and (4.29):

$$\dot{x}_k=g_k(x)=\delta_1^j a(x,b)+P(x,b)[(j-1)x_{j-1}-jx_j]$$

$$+\sum_{j<k}C_{j,k-j}(x,b)x_jx_{k-j}-2\sum_j C_{kj}(x,b)x_jx_k+2\sum_{j>k}F_{jk}(x,b)x_j-\sum_{j<k}F_{kj}(x,b)x_k.$$

$$(4.43)$$

The proof that this system does in fact represent the dynamic LLN for the Markov chains defined by the generator (4.42) (i.e., the proof of the convergence of these Markov chains as $h\to 0$) requires advanced tools from the theory of infinite-dimensional ODEs. It will be discussed in the next section under very general assumption of unbounded rates, namely under the assumption of the so-called *additive bounds for rates*:

$$a(x, b) + P(x, b) \leq c, \quad C_{kj}(x, b) \leq c(k + j), \quad \sum_{j<k} F_{kj}(x, b) \leq cj \quad (4.44)$$

with some constant c.

An important issue that we will not touch upon here is to understand the controllability of the limiting (now in the sense $t \to \infty$) stationary solutions, which may lead to the possibility of developing tools for influencing the power tails of distributions (Zipf's law) appearing in many situations of practical interest, as well as the proliferation or extinction of certain desirable (or undesirable) characteristics of the processes of evolution.

4.6 Unbounded Coefficients and Growing Populations

This section goes much further in mathematical sophistication. We shall exploit the theory of regularity and sensitivity of ODEs in l_1 with unbounded coefficients. The key ingredient in our argument is the following well-posedness and sensitivity result for the kinetic equation (4.43).

Theorem 4.6.1. *For some b, let $a(x, b)$, $P(x, b)$, $C(x, b)$, $F(x, b)$ be twice continuously differentiable in x and satisfy the additive bound condition (4.44) together with their derivatives, that is, (4.44) and the estimates*

$$\left|\frac{\partial a(x, b)}{\partial x_p}\right| + \left|\frac{\partial P(x, b)}{\partial x_p}\right| \leq c, \quad \left|\frac{\partial C_{kj}(x, b)}{\partial x_p}\right| \leq c(k + j), \quad \sum_{j<k}\left|\frac{\partial F_{kj}(x, b)}{\partial x_p}\right| \leq cj,$$

$$(4.45)$$

$$\left|\frac{\partial a^2(x, b)}{\partial x_p \partial x_q}\right| + \left|\frac{\partial^2 P(x, b)}{\partial x_p \partial x_q}\right| \leq c, \quad \left|\frac{\partial^2 C_{kj}(x, b)}{\partial x_p \partial x_q}\right| \leq c(k + j), \quad \sum_{j<k}\left|\frac{\partial^2 F_{kj}(x, b)}{\partial x_p \partial x_q}\right| \leq cj,$$

$$(4.46)$$

uniformly for all x, p, q.

Then for all $x \in \mathcal{M}^+(L^3)$, with the function $L_j = j$ there exists a unique global solution $X(t, x)$ of equation (4.43) that also belongs to $\mathcal{M}^+(L^3)$ for all times. Moreover, the derivatives with respect to the initial data

$$\xi^p(t, x) = \frac{\partial X(t, x)}{\partial x_p}, \quad \eta^{p,q}(t, x) = \frac{\partial^2 X(t, x)}{\partial x_p \partial x_q}$$

are well defined for all times and belong to the spaces $\mathcal{M}^+(L^2)$ and $\mathcal{M}^+(L)$ respectively, where they are uniformly bounded on bounded times and bounded x:

$$\sup_x \sum_j j|\xi_j^p(t, x)| \leq C(\lambda, T)L_p, \quad \sup_x \sum_j j^2|\xi_j^p(t, x)| \leq C(\lambda, T)L_p^2, \quad (4.47)$$

$$\sup_x \sum_j j |\eta_j^{p,q}(t,x)| \le t C(\lambda, T) L_p L_q, \tag{4.48}$$

uniformly for $t \le T$, where $\sup_x$ is over $x \in \mathcal{M}_{\le \lambda}(L^3)$, that is, over x satisfying the estimates $\sum_j j^3 x_j \le \lambda$.

This theorem is a slight extension of the general sensitivity result from [136]. Its full and detailed proof is given in Chapter 3 of [142].

We can now follow the line of argumentation from Theorem 2.4.1 or 4.3.1 to prove the convergence of the Markov chains $X_h(t, x(h))$ generated by (4.42), that is, to obtain the following main result of this chapter. Let U_h^t be the semigroup of the Markov chain $X_h(t, x(h))$, and let U^t be the semigroup generated by equation (4.43), that is, $U^t G(x) = G(X(t, x))$.

Theorem 4.6.2. *Under the assumption of Theorem 4.6.1, let the initial states $x(h)$ of the Markov chains $X_h(t, x(h))$ generated by (4.42) belong to $\mathcal{M}_{\le \nu}^+(L^3)$ with some ν and converge to a state x so that $\sum_j j |x_j(h) - x_j| \to 0$ as $h \to 0$. Then the Markov chains $X_h(t, x(h))$ converge in distribution to the deterministic evolution $X(t, x)$, and moreover, for all $G \in C^2(l_1)$ and $T > 0$,*

$$\sup_{t \le T} \|U_h^t G(x(h)) - U^t G(x)\| \le C_T (th + \sum_j j |x_j(h) - x_j|) \|G\|_{C^2(l_1)}. \tag{4.49}$$

Proof. Recall that

$$\|G\|_{C^2(l_1)} = \sup |G(x)| + \sup_{x,j} \left| \frac{\partial G}{\partial x_j} \right| + \sup_{x,j,k} \left| \frac{\partial^2 G}{\partial x_j \partial x_k} \right|.$$

Following the argument of Theorem 2.4.1 or 4.3.1, we have to obtain a bound for the expression $|(L_{b,h} - \Lambda_b) U^t G(x)|$ that is uniform for $x \in \mathcal{M}_{\le \nu}^+(L^3)$. Looking at the expressions for $L_{b,h}$ and Λ_b, we find that

$$\|(L_{b,h} - \Lambda_b) U^t G(x)\| = ha(x) \frac{\partial^2 \phi}{\partial x_1^2} + \frac{1}{2} h P(x) \sum_k k x_k \left(\frac{\partial^2 \phi}{\partial x_k^2} + \frac{\partial^2 \phi}{\partial x_{k+1}^2} - 2 \frac{\partial^2 \phi}{\partial x_k \partial x_{k+1}} \right)$$

$$+ \frac{1}{2} h \sum_{j,k} \left(\frac{\partial^2 \phi}{\partial x_{k+j}^2} + \frac{\partial^2 \phi}{\partial x_j^2} + \frac{\partial^2 \phi}{\partial x_k^2} + 2 \frac{\partial^2 \phi}{\partial x_k \partial x_j} - 2 \frac{\partial^2 \phi}{\partial x_k \partial x_{k+j}} - 2 \frac{\partial^2 \phi}{\partial x_j \partial x_{k+j}} \right)$$

$$+ \frac{1}{2} h \sum_{j,k} \left(\frac{\partial^2 \phi}{\partial x_{j-k}^2} + \frac{\partial^2 \phi}{\partial x_j^2} + \frac{\partial^2 \phi}{\partial x_k^2} - 2 \frac{\partial^2 \phi}{\partial x_k \partial x_j} + 2 \frac{\partial^2 \phi}{\partial x_k \partial x_{j-k}} - 2 \frac{\partial^2 \phi}{\partial x_j \partial x_{j-k}} \right),$$

where the derivatives are taken at points in small neighborhoods of x and where $\phi(x) = U^t G(x) = G(X(t, x))$, so that

$$\frac{\partial \phi}{\partial x_j} = \sum_i \frac{\partial G}{\partial x_i}(y)|_{y=X(t,x)}\xi_i^j(t,x),$$

$$\frac{\partial^2 \phi}{\partial x_p \partial x_q} = \sum_i \frac{\partial G}{\partial x_i}(y)|_{y=X(t,x)}\eta_i^{p,q}(t,x) + \sum_{i,m} \frac{\partial^2 G}{\partial x_i \partial x_m}(y)|_{y=X(t,x)}\xi_i^p(t,x)\xi_m^q(t,x).$$

Using the estimate for the derivatives of $X(t,x)$ from Theorem 4.6.1, all terms here are estimated in magnitude straightforwardly by some constants. For instance, let us look at the second term. It can be estimated as

$$P(x)\sum_k kx_k \left| \frac{\partial^2 \phi}{\partial x_k^2} + \frac{\partial^2 \phi}{\partial x_{k+1}^2} - 2\frac{\partial^2 \phi}{\partial x_k \partial x_{k+1}} \right|$$

$$\leq 2c\|G\|_{C^2(l_1)} \sum_k kx_k \sum_{i,j} \left(|\xi_i^k(t,x)\xi_j^k(t,x)| + |\xi_i^{k+1}(t,x)\xi_j^{k+1}(t,x)| + |\xi_i^k(t,x)\xi_j^{k+1}(t,x)| \right)$$

$$+2c\|G\|_{C^2(l_1)} \sum_k kx_k \sum_i \left(|\eta_i^{k,k}(t,x)| + |\eta_i^{k+1,k+1}(t,x)| + |\eta_i^{k,k+1}(t,x)| \right)$$

$$\leq C\|G\|_{C^2(l_1)} \sum_k kx_k(k^2 + (k+1)^2 + k(k+1)) \leq C\|G\|_{C^2(l_1)},$$

as required, with C depending on $\sum_k k^3 x_k$. Estimating analogously other terms, we get the required uniform bound for $|(L_{b,h} - \Lambda_b)U^t G(x)|$ and then complete the proof as in Theorem 4.3.1.   □

Analogously to Section 4.3 and using the theory of stability of the kinetic equations, we can generalize this result to the case of Lipschitz coefficients.

Theorem 4.6.3. *Let $P(x), a(x), Q_{kl}(x), P^{m,k}(x)$ be continuous nonnegative functions on l_1 such that*

$$P(x) + a(x) \leq c, \quad C_{kl}(x) \leq c(k+l), \quad \sum_{j<k} F^{kj}(x) \leq ck, \qquad (4.50)$$

$$|P(x) - P(y)| + |a(x) - a(y)| \leq c(L, |x-y|), \quad |C_{kl}(x) - C_{kl}(y)| \leq c(k+l)(L, |x-y|),$$
$$(4.51)$$

$$\sum_{j<k} |F^{kj}(x) - F^{kj}(y)| \leq ck(L, |x-y|), \qquad (4.52)$$

for $L = (1, 2, \cdots)$ and a constant c. Let the initial states $x(h)$ of the Markov chains $X_h(t, x(h))$ generated by (4.42) belong to $M_{\leq \nu}^+(L^3)$ with some ν and converge to a state x so that $\sum_j j|x_j(h) - x_j| \to 0$ as $h \to 0$. Then the Markov chains $X_h(t, x(h))$

converge in distribution to the deterministic evolution $X(t, x)$, and moreover, for every Lipschitz continuous G on l_1 and all $T > 0$,

$$\sup_{t \leq T} \|U_h^t G(x(h)) - U^t G(x(h))\| \leq C_T (th)^{2/5} \|G\|_{Lip}. \qquad (4.53)$$

Proof. The difference with the proof of Theorem 4.3.2 is that here, it is more convenient to approximate the coefficients $P(x), a(x), Q_{kl}(x), P^{m,k}(x)$ by their finite-dimensional approximations $P(\mathcal{P}_n(x)), a(\mathcal{P}_n(x)), Q_{kl}(\mathcal{P}_n(x)), P^{m,k}(\mathcal{P}_n(x))$, rather than the whole r.h.s. Let us denote the corresponding solution of the kinetic equations $X^n(t, x)$ and the corresponding approximating Markov chain by $X_h^n(t, x)$. As proved in Chapter 3 of [142], the evolutions $X^n(t, x)$ and $X(t, x)$ are well defined, and the deviation of the transition operators of the evolutions $X^n(t, x)$ and $X(t, x)$ is of order $1/n^2$ for the evolutions with the initial conditions from $\mathcal{M}_{\leq \nu}^+(L^3)$ with any given ν. Similarly, one shows the same deviation for the transition operators of the Markov chains $X_h^n(t, x)$ and $X_h(t, x)$. Approximating now the coefficients $P(\mathcal{P}_n(x)), a(\mathcal{P}_n(x)), Q_{kl}(\mathcal{P}_n(x)), P^{m,k}(\mathcal{P}_n(x))$ by the smooth functions $P_\delta(\mathcal{P}_n(x)), a_\delta(\mathcal{P}_n(x)), Q_{kl}^\delta(\mathcal{P}_n(x)), P_\delta^{m,k}(\mathcal{P}_n(x))$ as in Theorem 4.3.2, we get new approximations with the deviations from the transitions without δ of order δ. The required second derivative needed to apply Theorem 4.6.2 to these last approximations is of order n/δ again, as in Theorem 4.3.2. Thus the total error term is of order

$$\frac{ht}{\delta} + n\delta + \frac{1}{n^2}.$$

Choosing n and δ that make all terms of equal decay in ht yields the required order $(th)^{2/5}$.   □

So far in this chapter we kept the parameter b of the principal in all our results but did not use it. However, in this chapter we extended the convergence results of Chapter 2 devoted to finite-state models to the case of a countable state space. Once this is done, the results on the forward-looking principal of Chapter 3, where the parameter b becomes an important controlled variable, can be extended more or less automatically to this new setting of countable spaces, as in Theorem 4.3.3 of Section 4.3.

Part II
Pools of Rational Optimizers

Chapter 5
MFGs for Finite-State Models

In this chapter, we introduce the MFG framework for discrete state spaces, stressing the points that are most relevant for the models studied in the next chapters. As was already mentioned, the difference between the present MFG setting and the modeling of Part I is that now the small agents themselves become rational optimizers and are not supposed just to follow some prescribed deterministic or stochastic strategies (such as myopic behavior). For simplicity, in treating MFG we will exclusively use finite-state models and will not touch the extensions to countable state spaces, though the results of Chapter 4 allow one to extend large portions of the theory more or less directly to this more general framework.

5.1 Preliminaries: Controlled Markov Chains

Let us recall briefly the theory of controlled Markov chains (see details, e.g., in [113]). Let $Q = \{Q_{ij}(t, u)\}$ be a family of Q-matrices, $i, j \in \{1, \cdots, d\}$, depending continuously on a parameter $u \in U \subset \mathbf{R}^n$ and piecewise continuously on time t. By a (Markov) strategy we mean any collection of piecewise continuous functions $\hat{u}_j(t)$. A controlled Markov chain specified by such a strategy is the Markov chain $X_{s,i}^{\hat{u}}(t)$ with the Q-matrices $Q_{ij}(t, \hat{u}_i(t))$. Suppose one gets a payoff $J(t, j, u)$ per unit of time staying at j with control u around time t and the terminal payoff $S_T(j)$, paid at time T if the process terminates in j at time T. The corresponding *Markov control problem* is to find the maximal total payoff

$$S(t, j) = \max_{\hat{u}} \mathbf{E}[\int_t^T J(s, j(s), \hat{u}_{j(s)}(s))ds + S_T(X_{t,x}^{\hat{u}}(T))],$$

© Springer Nature Switzerland AG 2019
V. N. Kolokoltsov and O. A. Malafeyev, *Many Agent Games in Socio-economic Systems: Corruption, Inspection, Coalition Building, Network Growth, Security*, Springer Series in Operations Research and Financial Engineering, https://doi.org/10.1007/978-3-030-12371-0_5

where $j(s) = X^{\hat{u}}_{t,j}(s)$, and a strategy, called an *optimal strategy*, where this maximum is attained.

It is known that $S(t, j)$ solves the backward (evolving from the terminal time T backward) *Bellman equation*:

$$\frac{\partial S(t, j)}{\partial t} + \max_u [J(t, j, u) + \sum_{k=1}^{d} Q_{jk}(t, u) S(t, k)] = 0, \qquad (5.1)$$

with the terminal condition $S(T, j) = S_T^j$.

The standard heuristic derivation of this equation goes as follows. Assuming that S is smooth and taking into account that our Markov chain can have only one jump during a small period of time τ (see (2.3) for the approximations by discrete-time Markov chains), we can write approximately

$$S(t, j) = \max_u [J(t, j, u)\tau + \tau|Q_{jj}| \sum_{k \neq j} \frac{Q_{jk}}{|Q_{jj}|} S(t + \tau, k) + (1 - \tau|Q_{jj}|) S(t + \tau, j)].$$

This way of expressing the optimal payoff at time t via its values at future times is often referred to in general as the *principle of optimality*. Expanding S in a Taylor series and keeping only terms linear in τ (free terms $S(t, j)$ cancel) yields (5.1).

A rigorous justification is usually performed by first proving the well-posedness of the Bellman equation and then showing that its solution supplies the required maximum (the so-called *verification theorem* arguments).

5.2   MFG Forward–Backward Systems and the Master Equation

Suppose we have a family $Q = \{Q_{ij}(x, u)\}$ of Q-matrices, $i, j \in \{1, \cdots, d\}$, depending continuously on a parameter $u \in U$ and Lipschitz continuously on $x \in \Sigma_d$. Suppose there are N players, each moving according to Q and aiming at maximizing the payoff

$$\mathbf{E}[\int_t^T J(s, j(s), x(s), u(s))\, ds + S_T(j(T))], \qquad (5.2)$$

where $j(s)$ is the position of the player at time s. Here the motions of all players are coupled, since all transitions depend on the total distribution of players $x = (n_1, \cdots, n_d)/N$, where n_j denotes the number of players in state j at any given time.

The *MFG methodology* suggests the following solution concept for this problem for large N. Let an evolution of the distributions $x(t)$, $t \in [0, T]$, be a known continuous curve. Then every player should search for the maximal payoff

$$S(t, j) = \max_{\hat{u}} \mathbf{E}[\int_t^T J(s, j(s), x(s), \hat{u}_{j(s)}(s)) \, ds + S_T(j(T))],$$

where $j(s) = X_{s,j}^{\hat{u}}(s)$ is the Markov chain with the Q-matrices $Q_{ij}(x(t), \hat{u}_i(t))$. As follows from (5.1), $S(t, i)$ should satisfy the backward Bellman equation

$$\frac{\partial S(t, j)}{\partial t} + \max_u [J(t, j, x(t), u) + \sum_k Q_{jk}(x(t), u) S(t, k)] = 0, \qquad (5.3)$$

with the terminal condition $S(T, j) = S_T(j)$. Assume that we can find a solution $S(t, x)$ and the corresponding optimal strategy $\hat{u}_j(t) = \hat{u}_j(t, x(t))$ providing max in this equation at any time t. Now, if all players are using this optimal strategy, the mean-field interacting particle systems of N players given by the Q-matrices

$$\hat{Q}_{ij}(t, x) = Q_{ij}(x, \hat{u}_i(t))$$

converge, according to Theorem 2.5.1, to the solutions $X_{0,x(0)}(t)$ of the system of (forward) kinetic equations (2.40):

$$\dot{x}_k = \sum_{i \neq k} [x_i Q_{ik}(x, \hat{u}_i(t))) - x_k Q_{ki}(x, \hat{u}_k(t)))] = \sum_{i=1}^d x_i Q_{ik}(x, \hat{u}_i(t))), \quad k = 1, ..., d.$$

$$(5.4)$$

Let $\hat{x}(t)$ be the solution of this system with the initial condition $\hat{x}(0) = x(0)$. The consistency between the controlled dynamics and the mean-field evolution can be naturally described by the requirement that $\hat{x}(t) = x(t)$. This is exactly the forward–backward *MFG consistency problem*, or MFG consistency condition, also referred to by some authors as a *Nash–MFG equilibrium*. Equivalently, starting with a control $u_j(t)$, we can solve the corresponding kinetic equation to find the distribution $x(t)$ and then find the corresponding optimal control $\hat{u}_j(t)$ fitting the solution of the HJB (5.3). The MFG consistency condition can then be expressed by the equation $u(t) = \hat{u}(t)$. In any case, it can be expressed by saying that the pair $(\hat{x}(t), \hat{u}_j(t))$ provides a solution to the coupled *forward–backward system* (5.3)–(5.4), more precisely, to its *initial–terminal value problem* (initial x_0 is given for (5.4) and terminal S_T for (5.3)).

One can expect that solutions to the MFG consistency problem should provide some approximations to the solutions of games with N players in which each player is trying to maximize (5.2). Thus we are led to two basic problems of MFG theory: describe the solutions to the MFG consistency problem (say, prove an existence and/or uniqueness theorem) and provide an exact link between these solutions and the corresponding original game with a finite number of players. The latter task can be discussed with two approaches (often requiring different techniques): showing that Nash equilibria of games with N players converge to a solution of the MFG consistency problem, or showing that the solutions of the MFG consistency problem yield approximate Nash equilibria for finite-player games. In this book we will work with models in which the solutions can be calculated explicitly, and thus we shall not touch here upon the first problem of MFG theory mentioned above. The second problem will be addressed in the next section.

Though we will work with this forward–backward formulation, let us mention, for completeness, an alternative approach to the MFG consistency problem that arises from looking directly at the limiting evolution of the pair $(j(t), x(t))$, where $x(t)$ evolves according to the kinetic equations,

$$\dot{x}_k = \sum_{i=1}^{d} x_i Q_{ik}(x(t), \hat{u}_i(t)), \quad k = 1, ..., d,$$

and $j(t) = X^{\hat{u}}_{s,x}(t)$ is the Markov chain with the Q-matrices $Q_{ij}(x(t), \hat{u}_i(t))$, as in the controlled Markov process in the continuous state space $\{1, \cdots, d\} \times \Sigma_d$. This is, strictly speaking, no longer a chain, since it evolves by jumps and continuous displacements in the continuous state space. Nevertheless, in controlling the process with the objective to maximize (5.2), we find for the optimal payoff the Bellman equation in the form

$$\frac{\partial S(t, j, x)}{\partial t} + \max_{u_1, \cdots, u_d} \left[J(t, j, x, u_j) + \sum_k Q_{jk}(x, u_j)S(t, k, x) + \sum_{k,i} \frac{\partial S(t, j, x)}{\partial x_k} x_i Q_{ik}(x, u_i) \right] = 0.$$

$$(5.5)$$

This is obtained analogously to (5.1) from the approximate equation

$$S(t, j, x) = \max_{u_1, \cdots, u_d} [\tau J(t, j, x, u_j) + \tau |Q_{jj}| \sum_{k \neq j} \frac{Q_{jk}(x, u_j)}{|Q_{jj}|} S(t + \tau, k, X_{x,t}(t + \tau))$$

$$+ (1 - \tau |Q_{jj}|)S(t + \tau, j, X_{x,t}(t + \tau))],$$

or, discarding the higher-order terms in τ,

$$S(t, j, x) = \max_{u_1, \cdots, u_d} [J(t, j, x, u)\tau + \tau \sum_{k \neq j} Q_{jk}S(t, k, x) + (1 - \tau |Q_{jj}|)S(t + \tau, j, X_{x,t}(t + \tau))],$$

using the first-order Taylor approximation

$$S(t + \tau, j, X_{x,t}(t + \tau)) = S(t, j, x) + \tau \frac{\partial S}{\partial t}(t, j, x) + \tau \sum_{k,i} \frac{\partial S}{\partial x_k}(t, j, x) x_i Q_{ik}(x, u_i).$$

Equation (5.5) is called the *master equation* (in backward form). The next statement shows that this equation provides (at least when it is reasonably well posed) an alternative approach to the analysis of the MFG consistency problem, which selects the most effective solutions to backward–forward systems, thus forming an envelope for various solutions of the MFG consistency problem.

Proposition 5.2.1. *Let $S(t, j, x)$ be a smooth function solving equation (5.5) with the terminal condition $S_T(j)$ and giving the optimal payoff in the Markov decision problem on $\{1, \cdots, d\} \times \Sigma_d$ used to derive this equation. Let it be possible to choose a Lipschitz continuous (in x) selector $\tilde{u}_j(t, x)$ giving a maximum in (5.5) (for this solution S) and hence to build the corresponding trajectories $X_{x_0}(t)$ solving the kinetic equations*

$$\dot{x}_j = \sum_i x_i Q_{ij}(x, \tilde{u}_i(t, x))$$

with any initial x_0.

Then the pair $(X_{x_0}(t), \tilde{u}_j(t, X_{x_0}(t))$ is a solution to the MFG consistency problem, and

$$\tilde{S}(t, j) \le S(t, j, x_0), \tag{5.6}$$

for the payoff $\tilde{S}(t, j)$ on every other solution $(\hat{x}(t), \hat{u}_j(t))$ to the forward–backward MFG consistency problem with $\hat{x}(t) = X_{x_0}(t)$.

Proof. By the definition of $S(t, j, x)$ as the solution to a Markov decision problem in $\{1, \cdots, d\} \times \Sigma_d$, $S(t, j, x_0)$ is not less than the payoff that can be obtained by any player using any symmetric strategy (the same as all other players) given the dynamics $X_{x_0}(t)$ of the total distributions, implying (5.6). It follows that the payoff $S(t, j, x_0)$ cannot be improved by changing the strategies inside the class of symmetric strategies. Consequently, $\tilde{u}_i(t, x(t))$ provides the maximum payoff in the class of these strategies and hence provides a solution to the forward–backward MFG consistency problem. □

Some comments are in order here. MFG problems were introduced by Lasry–Lions [159] and Huang–Malhame–Caines [118], [119]. The deep theory developed so far is reflected in several books and surveys. We give a very brief bibliographical review of its development in Appendix. We were intentionally brief in our presentation, aiming at possibly the most elementary exposition needed to grasp the simplest finite-state models dealt with in the present book. For instance, as we have already mentioned, we will not use the master equation. We introduced it to give an idea of the major direction of research in MFG. In fact, the exploitation of the master equation is the only way to include in the theory the common noise and/or a major player, whose coupling with small players includes such noise (see [160]). The master equation is the limiting equation for the system of Bellman equations describing N-player games. In this sense, it is a decoupling equation encoding the entire mean-field game in one evolution. In particular, having a solution to the master equation leads to a solution of the convergence problem. All these aspects are fully discussed in [56] and the fundamental two-volume monograph [60]. In [160], [56], [60], one can also find many deep results on the existence and/or uniqueness of the master equation, including some cases for small players with a finite state space.

5.3　MFG Solutions and Nash Equilibria for a Finite Number of Players

In the next chapters we shall construct the solutions to the MFG consistency problem for several models and discuss the properties of these solutions, without paying attention to their links with the related games of finitely many players. The justification for this study will be provided now, where we shall show that these solutions represent approximate Nash equilibria for N-player games. The results of this section will not be used explicitly in our further analysis.

Let us recall that a *Nash equilibrium* in a game of N players is a collection of strategies of these players, often referred to as a profile of strategies, such that a unilateral deviation of any particular player from this profile cannot improve the payoff of this agent. Therefore, as Nash himself stressed, this is a no-regret outcome for each player. An ϵ-*Nash equilibrium* is a profile of strategies such that a unilateral deviation of any particular player from this profile can improve the payoff of this agent by an amount not exceeding ϵ.

Recall that our functional norms always refer to the dependence on x uniform with respect to other variables, so that, for instance,

$$\|Q\|_{C^1} = \|Q\| + \sup_i \sum_j \left| \sup_{k,x,u} \frac{\partial Q_{ij}}{\partial x_k} \right|, \quad \|Q\|_{C^2} = \|Q\|_{C^1} + \sup_i \sum_j \left| \sup_{k,l,x,u} \frac{\partial^2 Q_{ij}}{\partial x_k \partial x_l} \right|.$$

The Nash equilibria and the ϵ-Nash equilibria for dynamic N-player games can be looked at in several ways, which are traditionally distinguished in the literature on optimization theory. In general, one speaks about *open loop control* and related *open loop equilibria* if players choose their control strategies $u_j(t)$ from the beginning, irrespective of the dynamics of the game (but which may depend on the common source of uncertainty). One speaks about *closed loop control* and related *closed loop equilibria* if players choose *feedback controls* $u_j(t, z)$, which at any time t depend also on the position z of the process. In the MFG setting, new possibilities arise, since from the point of view of each player, the position includes the player's own position, say j, and the overall distribution x. Let us speak about *partially open loop control* and related *partially open loop equilibria* if each player chooses among strategies $u_j(t)$, piecewise continuous in t, that depend on the player's own position at time t, but not on the overall distribution x. The use of such strategies, sometimes referred to as *distributed strategies*, is reasonable in many cases in which the overall distribution is not easily observable by each concrete player. By a *closed loop control* we mean, as usual, a control $u_j(t, x)$ that is supposed to be applied by a player at time t when its position is j and the overall distribution is x.

For the sake of transparency, we shall concentrate on the model of Section 5.2 with the running cost function J in (5.2) not depending on x explicitly, that is, with the payoff

$$\mathbf{E}[\int_t^T J(s, j(s), u_{j(s)}(s))\, ds + S_T(j(T))], \tag{5.7}$$

with a continuous function J, and with partially open loop equilibria. Thus we assume that we have a family $Q = \{Q_{ij}(x, u)\}$ of Q-matrices, $i, j \in \{1, \cdots, d\}$, depending continuously on a parameter $u \in U$ and Lipschitz continuously on $x \in \Sigma_d$. Suppose there are N players, each moving according to Q and aiming at maximizing the payoff (5.7).

Theorem 5.3.1. *Let $(\hat{x}(t), \hat{u}_j(t))$ be a solution to the backward–forward MFG consistency problem. Then for the initial distribution $x(0)$, the symmetric profile of strategies $\hat{u}_j(t))$ is an ϵ-Nash equilibrium in the partially open loop setting, with ϵ of order $1/\sqrt{N}$. If $Q(x, u)$ is twice continuously differentiable in x uniformly in u, then the order of ϵ improves to $1/N$.*

Proof. We have to show that if all players use the strategy $\hat{u}_j(t)$, then any particular player unilaterally deviating from this strategy cannot increase the payoff by an amount exceeding ϵ. Thus let us assume that one tagged player is using some deviating strategy $u_j^{dev}(t)$, while other players stick to $\hat{u}_j(t)$. The natural state space for such a Markov chain will be $\{1, \cdots, d\} \times \Sigma_d$, the first discrete coordinate j denoting the position of the tagged player.

We are exactly in the setting of Section 2.7. The operator $L_t^{N,dev}$ given by (2.79) and the limiting operator (2.80 take the form

$$L_t^{N,dev} f(j, x) = \sum_k Q_{jk}^{dev}(x, u_j^{dev}(t))(f(k, x) - f(j, x))$$

$$+ \sum_i (x_i - \delta_i^j/N) \sum_{k \neq i} Q_{ik}(x, \hat{u}_j(t))\,[f(j, x - e_i/N + e_k/N) - f(j, x)], \tag{5.8}$$

$$\Lambda_t^{dev} f(j, x) = \sum_k Q_{jk}^{dev}(x, u_j^{dev}(t))(f(k, x) - f(j, x))$$

$$+ \sum_i x_i \sum_{k \neq i} Q_{ik}(x, \hat{u}_j(t)) \left[\frac{\partial f}{\partial x_k} - \frac{\partial f}{\partial x_i} \right] (j, x). \tag{5.9}$$

Applying Theorem 2.7.1, we derive that the payoffs for the tagged player in the N-player game differ by that player's payoff in the limiting evolution by an amount not exceeding ϵ. Notice that in order to take into account the running payoff J, we apply this theorem not only to $S_T(j)$ and terminal time T, but also to each $J(s, j(s), u_{j(s)}(s))$ with terminal time s. Since $\hat{u}_j(t)$ is optimal in the limiting game, it is therefore ϵ-optimal for the N-player game. $\square$

Remark 20. *To extend the theorem to the closed-loop control and J depending on x, one just has to apply Theorem 2.7.2, which extends Theorem 2.7.1 to functions f depending explicitly on x.*

5.4   MFG with a Major Player

We complete here our brief review of the foundations of MFGs on finite state spaces by explaining two popular extensions of the basic model, which includes a major player. Namely, let us assume, as in Sections 2.6 and 3.3, that the transition matrices $Q = Q(x, u, b)$ and the payoffs $J(t, j, x, u, b)$ depend additionally on a parameter b controlled by the major player (a principal). If, as in Section 2.6, the principal is playing just the best response $b^*(x)$, we are directly back in the original problem with $Q = Q(x, u, b^*(x))$. However, if the major player chooses b strategically, aiming at maximizing some payoff of the general type

$$\int_t^T B(x(s), b(s)) \, ds + V_T(x(T)),$$

the situation becomes different.

In this case, the *MFG methodology* works as follows. If the evolution of the distributions $x(t)$, $t \in [0, T]$, is a known continuous curve, then the major player finds the optimal strategy $\hat{b}(t)$ and, based on this strategy, any given player should search for the maximal payoff

$$S(t, i) = \max_{\hat{u}} \mathbf{E}\left[\int_t^T J(s, j(s), x(s), \hat{u}_{j(s)}(s), \hat{b}(s)) dt + S_T(j(T))\right],$$

where $j(t) = X_{s,x}^{\hat{u}}(t)$ is the Markov chain with Q-matrices $Q_{ij}(x(t), \hat{u}_i(t), \hat{b}(t))$. The Bellman equation for the optimal payoff of each small player $S(t, i)$ takes the form

$$\frac{\partial S(t, j)}{\partial t} + \max_u [J(t, j, x(t), u, \hat{b}(t)) + \sum_k Q_{jk}(x(t), u, b(t)) S(t, k)] = 0.$$

(5.10)

After finding a solution $S(t, x)$ and the corresponding optimal strategy $\hat{u}_j(t) = \hat{u}_j(t, x(t))$ providing max in this equation at any time t, we can solve the corresponding kinetic equations

$$\dot{x}_k = \sum_{i=1}^d x_i Q_{ik}(x, \hat{u}_i(t)), \hat{b}(t)), \quad k = 1, ..., d. \tag{5.11}$$

Let $\hat{x}(t)$ be the solution of this system with the initial condition $\hat{x}(0) = x(0)$. The *MFG consistency problem*, or *MFG consistency condition with major player*, can be expressed by the equation $\hat{x}(t) = x(t)$.

Chapter 6
Three-State Model of Corruption and Inspection

A simple model of corruption that takes into account the effect of the interaction of a large number of agents by both rational decision-making and myopic behavior is developed. It describes the distribution of corrupt and honest agents under the pressure of both an incorruptible governmental representative (often referred to in the literature as a "benevolent principal"; see, e.g., [4]) and the "social norms" of the society. Its stationary version turns out to be a rare example of an exactly solvable model of mean-field-game type. In particular, it reveals explicitly the nonuniqueness of solutions, which is widely discussed in the general literature on mean-field game theory. The results show clearly how the presence of interactions (including social norms) influences the spread of corruption by creating certain phase transitions from one to three equilibria. The properties of this model motivate the study of a more general class of models in Chapter 8, but formally, those general results are independent of the exposition of this chapter.

The stability of stationary solutions is analyzed here only in a preliminary reduced sense, as the stability of the rest point of kinetic equations under fixed control. The proper stability in the sense of the full forward–backward system will be discussed in Chapter 8 under the setting of the discounted control problem leading to the turnpike property of these stationary solutions.

Let us stress finally that in all our MFG models, the major player is effectively left in the background (it is represented by an eternal parameter) and is not analyzed strategically. We consider the proper analysis of the full versions of these games with a major player (as is done in Chapter 3 for myopic small payers) as a promising direction for future research.

© Springer Nature Switzerland AG 2019

V. N. Kolokoltsov and O. A. Malafeyev, *Many Agent Games in Socio-economic Systems: Corruption, Inspection, Coalition Building, Network Growth, Security*, Springer Series in Operations Research and Financial Engineering, https://doi.org/10.1007/978-3-030-12371-0_6

6.1   The Model

Suppose an agent can be in one of three states: honest H, corrupt C, reserved R, where R is a reserved job of low salary that an agent receives as a punishment if its corrupted behavior is discovered.

The change between H and C is subject to the decisions of the agents (though the precise time of the execution of their intent is noisy); the changes from C to R are random with distributions depending on the level of the efforts (say, a budget used) b of the principal (a government representative) invested in chasing corrupt behavior; the change R to H (say, a new recruitment) may be possible and is included as a random event with a certain rate

Let n_H, n_C, n_R denote the numbers of agents in the corresponding states with $N = n_H + n_C + n_R$ the total number of agents. By a state of the system we shall mean either the 3-vector $n = (n_H, n_C, n_R)$ or its normalized version $x = (x_H, x_C, x_R) = n/N$.

The control parameter u of each player in state H or C may have two values, 0 and 1, meaning that the player is happy with its state (H or C) or it prefers to switch from one to the other; there is no control in the state R. When the updating decision 1 is made, the updating effectively occurs with some rate λ. The recovery rate, that is, the rate of change from R to H (we assume that once recruited, the agents start by being honest) is a given constant r.

Remark 21. *The choice of bang-bang controls with two values seems to be easiest to interpret: you are either happy with your state or want to change. The control $u \in (0, 1)$ would be more vague (and more difficult to evaluate) here. However, since u enters linearly in the HJB (see below), the maximizers would anyway belong to $\{0, 1\}$ even if $u \in [0, 1]$ were allowed.*

Apart from taking a rational decision to swap H and C, an honest agent can be pushed to become corrupt by its corrupt peers, the effect being proportional to the fraction of corrupt agents with a certain coefficient q_{inf}, which is analogous to the infection rate in epidemiological models. On the other hand, the honest agents can contribute to chasing and punishing corrupt behavior, this effect of a desirable social norm being proportional to the fraction of honest agents with a certain coefficient q_{soc}. The presence of the coefficients q_{inf}, q_{soc}, reflecting the social interaction, makes the dynamics of individual agents dependent on the distribution of other agents, thus bringing the model into the setting of mean-field games. It is our major concern to find out how the presence of interaction influences the spread of corruption.

Thus if all agents use the strategy $u_H, u_C \in \{0, 1\}$ and the efforts of the principal is b, then the evolution of the state x is clearly given by the ODE

$$\begin{cases} \dot{x}_R = (b + q_{soc}x_H)x_C - rx_R, \\ \dot{x}_H = rx_R - \lambda(x_H u_H - x_C u_C) - q_{inf}x_H x_C, \\ \dot{x}_C = -(b + q_{soc}x_H)x_C + \lambda(x_H u_H - x_C u_C) + q_{inf}x_H x_C. \end{cases} \tag{6.1}$$

Here u_H, u_C can be considered arbitrary measurable functions of t.

This system is a particular representative of kinetic equations (5.4). Let us write down explicitly the generator of the corresponding approximating Markov chain of N players, in order to fit the model fully and explicitly in the general framework of Section 5.2. If all agents use the strategy $u_H, u_C \in \{0, 1\}$ and the effort of the principal is b, the generator of the Markov evolution on the states n is

$$L_N F(n_H, n_C, n_R) = n_C(b + q_{soc}\frac{n_H}{N})(F(n_H, n_C - 1, n_R + 1) - F(n_H, n_C, n_R))$$

$$+n_R r(F(n_H+1, n_C, n_R - 1) = F(n_H, n_C, n_R)) + \lambda n_C u_C(F(n_H + 1, n_C - 1, n_R) - F(n_H, n_C, n_R))$$

$$+n_H(\lambda u_H + q_{inf}\frac{n_C}{N})(F(n_H - 1, n_C + 1, n_R) - F(n_H, n_C, n_R)).$$

For every N, this generator describes a Markov chain on the finite state space $\{n = (n_H, n_C, n_R) : n_H + n_C + n_R = N\}$, where any agent, independently of others, can be recruited with rate r (if in state R) or change from C to H or vice versa if desired (with rate λ), and where changes of state due to binary interactions are taken into account by the terms containing q_{soc} and q_{inf}.

In terms of x, the generator $L_N F$ takes the form

$$L_N F(x) = x_C(b + q_{soc}x_H)(F(x - e_C/N + e_R/N) - F(x))$$

$$+x_R r(F(x - e_R/N + e_H/N) - F(x)) + \lambda x_C u_C(F(x - e_C/N + e_H/N) - F(x))$$

$$+ x_H(\lambda u_H + q_{inf}x_C)(F(x - e_H/N + e_C/N) - F(x)), \tag{6.2}$$

where $\{e_j\}$ is the standard basis in $\mathbf{R}^3$.

If F is a differentiable function, then $L_N F$ converges to

$$LF(x) = x_C(b + q_{soc}x_H)\left(\frac{\partial F}{\partial x_R} - \frac{\partial F}{\partial x_C}\right) + x_R r\left(\frac{\partial F}{\partial x_H} - \frac{\partial F}{\partial x_R}\right)$$

$$+ x_H(\lambda u_H + q_{inf}x_C)\left(\frac{\partial F}{\partial x_C} - \frac{\partial F}{\partial x_H}\right) + \lambda x_C u_C\left(\frac{\partial F}{\partial x_H} - \frac{\partial F}{\partial x_C}\right) \tag{6.3}$$

as $N \to \infty$, which follows from Taylor's formula. This is a first-order partial differential operator, and its characteristics are given by the ODE (6.1).

If $x(t)$ and $b(t)$ are given, then the dynamics of each individual player is the Markov chain on the three states, with the generator

$$\begin{cases} L^{ind}g(R) = r(g(H) - g(R)) \\ L^{ind}g(H) = (\lambda u_H^{ind} + q_{inf}x_C)(g(C) - g(H)) \\ L^{ind}g(C) = \lambda u_C^{ind}(g(H) - g(C)) + (b + q_{soc}x_H)(g(R) - g(C)) \end{cases} \quad (6.4)$$

depending on the individual control $u^{ind} \in \{0, 1\}$, so that $\dot{g} = L^{ind}g$ is the Kolmogorov backward equation of this chain.

Assume that an employed agent receives a wage w_H per unit of time and, if corrupt, an average payoff w_C (which includes w_H plus some additional illegal reward); the agent has to pay a fine f if its illegal behavior is discovered; the reserved wage for fired agents is w_R. Thus the total payoff for a player in the time period $[t, T]$ is $\int_t^T w_S(\tau)d\tau + fM(t, T)$, where S denotes the state (which is either R, H, or C) and $M(t, T)$ is the number of transitions from C to R during the period. If the distribution of other players is $x(t) = (x_R, x_H, x_C)(t)$, then the HJB equation of type (5.5) describing the expectation of the optimal payoff $g = g_t$ (starting at time t with time horizon T) of an agent is

$$\begin{cases} \dot{g}(R) + w_R + r(g(H) - g(R)) = 0 \\ \dot{g}(H) + w_H + \max_u(\lambda u + q_{inf}x_C)(g(C) - g(H)) = 0 \\ \dot{g}(C) + w_C - (b + q_{soc}x_H)f + \max_u(\lambda u(g(H) - g(C)) + (b + q_{soc}x_H)(g(R) - g(C))) = 0. \end{cases} \quad (6.5)$$

Therefore, starting with some control

$$u^{com}(t) = (u_C^{com}(t), u_H^{com}(t))$$

used by all players, we can find the dynamics $x(t)$ from equation (6.1) (with u^{com} used for u). Then each individual should solve the Markov control problem (6.5) thus finding the individually optimal strategy

$$u^{ind}(t) = (u_C^{ind}(t), u_H^{ind}(t)).$$

The *MFG consistency condition* of Section 5.2 can now be explicitly written as

$$u^{ind}(t) = u^{com}(t). \quad (6.6)$$

6.2   Stationary MFG Consistency Problem

Instead of analyzing the complicated dynamic problem (6.6), we shall look for a simpler and practically more relevant problem of consistent stationary strategies.

There are two standard stationary problems arising from HJB (6.5), one being the search for the average payoff

$$g = \lim_{T \to \infty} \frac{1}{T} \int_0^T g_t \, dt$$

for long-period games, and the other a search for a discounted optimal payoff. The first is governed by the solutions of HJB of the form $(T - t)\mu + g$, linear in t (with μ describing the optimal average payoff), so that g satisfies the stationary HJB equation

$$\begin{cases} w_R + r(g(H) - g(R)) = \mu \\ w_H + \max_u(\lambda u + q_{inf} x_C)(g(C) - g(H)) = \mu \\ w_C - (b + q_{soc} x_H) f + \max_u(\lambda u(g(H) - g(C)) + (b + q_{soc} x_H)(g(R) - g(C)) = \mu, \end{cases}$$
$$(6.7)$$

and the discounted optimal payoff (with the discounting coefficient δ) satisfies the stationary HJB

$$\begin{cases} w_R + r(g(H) - g(R)) = \delta g(R) \\ w_H + \max_u(\lambda u + q_{inf} x_C)(g(C) - g(H)) = \delta g(H) \\ w_C - (b + q_{soc} x_H) f + \max_u(\lambda u(g(H) - g(C)) + (b + q_{soc} x_H)(g(R) - g(C)) = \delta g(C). \end{cases}$$
$$(6.8)$$

The analyses of these two settings are mostly analogous, since they are in some sense equivalen; see, e.g., [203] (they are analogous for constructing the MFG equilibria, but different for the analysis of precise links with a finite-N problem). For this model we shall concentrate on the first one.

For a fixed b, the *stationary MFG consistency problem* for the average payoff is to find $(x, u_C, u_H) = (x, u_C(x), u_H(x))$, where x is the stationary point of evolution (6.1), that is,

$$\begin{cases} (b + q_{soc} x_H) x_C - r x_R = 0 \\ r x_R - \lambda(x_H u_H(x) - x_C u_C(x)) - q_{inf} x_H x_C = 0 \\ -(b + q_{soc} x_H) x_C + \lambda(x_H u_H(x) - x_C u_C(x)) + q_{inf} x_H x_C = 0, \end{cases}$$
$$(6.9)$$

where $u_C(x)$, $u_H(x)$ are the maximizers in (6.7). Thus x is a fixed point of the limiting dynamics of the distribution of a large number of agents such that the corresponding stationary control is individually optimal subject to this distribution.

Fixed points can practically model a stationary behavior only if they are stable. Thus we are interested in stable solutions $(x, u_C, u_H) = (x, u_C(x), u_H(x))$ to the stationary MFG consistency problem, where a solution is stable if the corresponding stationary distribution $x = (x_R, x_H, x_C)$ is a stable equilibrium to (6.1) (with u_C, u_H fixed by this solution). By stability of a fixed point of a dynamics, we mean the usual dynamic stability: if the dynamics has started in a sufficiently small neighborhood of this point, then it converges to it as time tends to infinity. A fixed point is called unstable if in any its neighborhoods there are initial points such that the dynamics will not converge to the fixed point when starting from these points. As mentioned above, our major concern is to find out how the presence of interaction (specified by the coefficients q_{soc}, q_{inf}) affects the stable equilibria.

6.3   Solutions to the Stationary MFG Problem

Our first result describes explicitly all solutions to the stationary MFG consistency problem.

We shall say that in a solution to the stationary MFG consistency problem the optimal individual behavior is corrupt if $u_C = 0, u_H = 1$: if you are corrupt, you stay corrupt, and if you are honest, you begin exhibiting corrupt behavior as soon as possible; the optimal individual behavior is honesty if $u_C = 1, u_H = 0$: if you are honest, you stay honest; if you are involved in corruption, try to cleanse yourself from corruption as soon as possible.

The natural assumptions on our coefficients, arising from the interpretation above, are

$$\lambda > 0, r > 0, b > 0, \quad f \geq 0, q_{soc} \geq 0, q_{inf} \geq 0, \quad w_C > w_H > w_R \geq 0. \quad (6.10)$$

The key parameter for our model turns out to be the quantity

$$\bar{x} = \frac{1}{q_{soc}} \left[\frac{r(w_C - w_H)}{w_H - w_R + rf} - b \right]. \quad (6.11)$$

(which can take values $\pm\infty$ if $q_{soc} = 0$).

Theorem 6.3.1. *Assume* (6.10).

(i) If $\bar{x} > 1$, then there exists a unique solution $x^* = (x_R^*, x_C^*, x_H^*)$ to the stationary MFG problem (6.9), (6.7), *where*

$$x_C^* = \frac{(1 - x_H^*)r}{r + b + q_{soc}x_H^*} \quad (6.12)$$

and x_H^* is the unique solution on the interval $(0, 1)$ of the quadratic equation $Q(x_H) = 0$, where

$$Q(x_H) = [(r + \lambda)q_{soc} - rq_{inf}]x_H^2 + [r(q_{inf} - q_{soc}) + \lambda r + \lambda b + rb]x_H - rb. \quad (6.13)$$

Under this solution, the optimal individual behavior is corruption: $u_C = 0, u_H = 1$.

(ii) If $\bar{x} < 1$, there may be one, two, or three solutions to the stationary MFG problem (6.9), (6.7). *Namely, the point $x_H = 1, x_C = x_R = 0$ is always a solution, under which the optimal individual behavior is being honest: $u_C = 1, u_H = 0$.*

Moreover, if

$$\max(\bar{x}, 0) \leq \frac{b+\lambda}{q_{inf} - q_{soc}} < 1, \tag{6.14}$$

then there is another solution with the optimal individual behavior being honest, that is, $u_C = 1, u_H = 0$:

$$x_H^{**} = \frac{b+\lambda}{q_{inf} - q_{soc}}, \quad x_C^{**} = \frac{r(q_{inf} - q_{soc} - b - \lambda)}{(r+b)q_{inf} + (\lambda - r)q_{soc}}. \tag{6.15}$$

Finally, if

$$\bar{x} > 0, \quad Q(\bar{x}) \geq 0, \tag{6.16}$$

then there is a solution with the corrupt optimal behavior of the same structure as in (i), that is, with x_H^* the unique solution to $Q(x_H) = 0$ on $(0, \bar{x}]$ and x_C^* given by (6.12).

Remark 22. *As seen by inspection, if $q_{inf} - q_{soc} > 0$, then $Q[(b+\lambda)/(q_{inf} - q_{soc})] > 0$, so that for $\bar{x}$ slightly less than $x_H^{**} = (b+\lambda)/(q_{inf} - q_{soc})$, one has also $Q(\bar{x}) > 0$, in which case one really has three points of equilibrium given by $x_H^*, x_H^{**}, x_H = 1$, with $0 < x^* < \bar{x} < x^{**} < 1$.*

Remark 23. *In case of the stationary problem arising from the discounting payoff, which is from equation (6.8), the role of the classifying parameter $\bar{x}$ from (6.11) is played by the quantity*

$$\bar{x} = \frac{1}{q_{soc}} \left[\frac{(r+\delta)(w_C - w_H)}{w_H - w_R + (r+\delta)f} - b \right]. \tag{6.17}$$

Proof. Clearly solutions to (6.7) are defined up to an additive constant. Thus we may assume that $g(R) = 0$. Moreover, we can reduce the analysis to the case $w_R = 0$ by subtracting it from all equations of (6.7) and thus shifting by w_R the values w_H, w_C, μ. Under these simplifications, the first equation in (6.7) is $\mu = rg(H)$, so that (6.7) becomes the system

$$\begin{cases} w_H + \lambda \max(g(C) - g(H), 0) + q_{inf}x_C(g(C) - g(H)) = rg(H), \\ w_C - (b + q_{soc}x_H)f + \lambda \max(g(H) - g(C), 0) - (b + q_{soc}x_H)g(C) = rg(H), \end{cases} \tag{6.18}$$

for the pair $(g(H), g(C))$ with $\mu = rg(H)$.

Assuming $g(C) \geq g(H)$, that is, $u_C = 0, u_H = 1$, so that corrupt behavior is optimal, system (6.18) turns into

$$\begin{cases} w_H + \lambda(g(C) - g(H)) + q_{inf}x_C(g(C) - g(H)) = rg(H), \\ w_C - (b + q_{soc}x_H)f - (b + q_{soc}x_H)g(C) = rg(H). \end{cases} \tag{6.19}$$

Solving this system of two linear equations, we get

$$g(C) = \frac{(r + \lambda + q_{inf}x_C)[w_C - (b + q_{soc}x_H)f] - rw_H}{r(\lambda + q_{inf}x_C + b + q_{soc}x_H) + (\lambda + q_{inf}x_C)(b + q_{soc}x_H)},$$

$$g(H) = \frac{(\lambda + q_{inf}x_C)[w_C - (b + q_{soc}x_H)f] + (b + q_{soc}x_H)w_H}{r(\lambda + q_{inf}x_C + b + q_{soc}x_H) + (\lambda + q_{inf}x_C)(b + q_{soc}x_H)},$$

so that $g(C) \geq g(H)$ is equivalent to

$$w_C - (b + q_{soc}x_H)f \geq w_H \left(1 + \frac{b + q_{soc}x_H}{r}\right),$$

or in other words,

$$x_H \leq \frac{1}{q_{soc}}\left[\frac{r(w_C - w_H)}{w_H + rf} - b\right], \tag{6.20}$$

which by restoring w_R (shifting w_C, w_H by w_R) gives

$$x_H \leq \bar{x} = \frac{1}{q_{soc}}\left[\frac{r(w_C - w_H)}{w_H - w_R + rf} - b\right]. \tag{6.21}$$

Since $x_H \in (0, 1)$, this is automatically satisfied if $\bar{x} > 1$, that is, under the assumption of (i). On the other hand, it definitely cannot hold if $\bar{x} < 0$.

Assuming $g(C) \leq g(H)$, that is, $u_C = 1$, $u_H = 0$, so that honest behavior is optimal, system (6.18) turns into

$$\begin{cases} w_H + q_{inf}x_C(g(C) - g(H)) = rg(H), \\ w_C - (b + q_{soc}x_H)f + \lambda(g(H) - g(C)) - (b + q_{soc}x_H)g(C) = rg(H). \end{cases} \tag{6.22}$$

Solving this system of two linear equations, we get

$$g(C) = \frac{(r + q_{inf}x_C)[w_C - (b + q_{soc}x_H)f] + (\lambda - r)w_H}{r(\lambda + q_{inf}x_C + b + q_{soc}x_H) + q_{inf}x_C(b + q_{soc}x_H)}$$

$$g(H) = \frac{q_{inf}x_C[w_C - (b + q_{soc}x_H)f] + (\lambda + b + q_{soc}x_H)w_H}{r(\lambda + q_{inf}x_C + b + q_{soc}x_H) + q_{inf}x_C(b + q_{soc}x_H)},$$

so that $g(C) \leq g(H)$ is equivalent to the inverse of condition (6.20).

If $g(C) \geq g(H)$, that is, $u_C = 0$, $u_H = 1$, then the fixed-point equation (6.9) becomes

$$\begin{cases} (b + q_{soc}x_H)x_C - rx_R = 0, \\ rx_R - \lambda x_H - q_{inf}x_Hx_C = 0, \\ -(b + q_{soc}x_H)x_C + \lambda x_H + q_{inf}x_Hx_C = 0. \end{cases} \tag{6.23}$$

Since $x_R = 1 - x_H - x_C$, the third equation is a consequence of the first two equations, which yields the system

$$(b + q_{soc}x_H)x_C - r(1 - x_H - x_C) = 0,$$
$$r(1 - x_H - x_C) - \lambda x_H - q_{inf}x_H x_C = 0. \tag{6.24}$$

From the first equation we have

$$x_C = \frac{(1 - x_H)r}{r + b + q_{soc}x_H}. \tag{6.25}$$

From this, it is seen that if $x_H \in (0, 1)$ (as it should be), then also $x_C \in (0, 1)$ and

$$x_C + x_H = \frac{r + x_H(b + q_{soc}x_H)}{r + b + q_{soc}x_H} \in (0, 1).$$

Plugging x_C into the second equation of (6.24), we obtain for x_H the quadratic equation $Q(x_H) = 0$ with Q given by (6.13).

Since $Q(0) < 0$ and $Q(1) > 0$, the equation $Q(x_H) = 0$ has exactly one positive root $x_H^* \in (0, 1)$. Hence x_H^* satisfies (6.20) if and only if either $\bar{x} > 1$ (that is, we are under the assumption of (i)) or (6.16) holds, proving the last statement of (ii).

If $g(C) \leq g(H)$, that is, $u_C = 1$, $u_H = 0$, then the fixed-point equation (6.9) becomes

$$\begin{cases} (b + q_{soc}x_H)x_C - x_R r = 0, \\ x_R r + \lambda x_C - q_{inf}x_H x_C = 0, \\ -x_C(b + q_{soc}x_H) - \lambda x_C + q_{inf}x_H x_C = 0. \end{cases} \tag{6.26}$$

Again here $x_R = 1 - x_H - x_C$, and the third equation is a consequence of the first two equations, which yields the system

$$\begin{cases} (b + q_{soc}x_H)x_C - r(1 - x_H - x_C) = 0, \\ r(1 - x_H - x_C) + \lambda x_C - q_{inf}x_H x_C = 0. \end{cases} \tag{6.27}$$

From the first equation we again get (6.25). Plugging this x_C into the second equation of (6.24), we obtain the equation

$$r(1 - x_H) = (r - \lambda + q_{inf}x_H)\frac{(1 - x_H)r}{r + b + q_{inf}x_H},$$

with two explicit solutions yielding the first and the second statements of (ii). $\qquad \square$

6.4 Stability of Solutions

Theorem 6.4.1. *Assume* (6.10).

(i) *The solution* $x^* = (x_R^*, x_C^*, x_H^*)$ *(given by Theorem 6.3.1) with individually optimal behavior being corruption is stable if*

$$-\frac{\lambda q_{soc}}{r} \leq q_{soc} - q_{inf} \leq \frac{r q_{inf} + (r+b)(br + r\lambda + b\lambda)}{r^2}. \qquad (6.28)$$

(ii) *Suppose* $\bar{x} < 1$. *If* $q_{inf} - q_{soc} \leq 0$ *or*

$$q_{inf} - q_{soc} > 0, \quad \frac{b + \lambda}{q_{inf} - q_{soc}} > 1,$$

then $x_H = 1$ *is the unique stationary MFG solution with individually optimal strategy being honest; and this solution is stable. If* (6.14) *holds, then there are two stationary MFG solutions with the individually optimal strategy being honest, one with* $x_H = 1$ *and the other with* $x_H = x_H^{**}$ *given by* (6.15); *the first solution is unstable, and the second is stable.*

We are not presenting necessary and sufficient conditions for the stability of solutions with optimally corrupt behavior. Condition (6.28) is only sufficient, but it covers a reasonable range of parameters for which the "epidemic" spread of corruption and social cleansing are effects of comparable order.

As a trivial consequence of our theorems we can conclude that in the absence of interaction, that is, for $q_{inf} = q_{soc} = 0$, corruption is individually optimal if

$$w_C - w_R \geq bf + (w_H - w_R)(1 + b/r), \qquad (6.29)$$

and honesty is individually optimal otherwise (which is, of course, a reformulation of the standard result for the basic model of corruption; see, e.g., [4]). In the first case, the unique equilibrium is

$$x_H^* = \frac{rb}{\lambda r + \lambda b + rb}, \quad x_C^* = \frac{r(1 - x_H^*)}{r + b}, \qquad (6.30)$$

and in the second case, the unique equilibrium is $x_H = 1$. Both are stable.

Proof. Notice that $(d/dt)(x_R + x_H + x_C) = 0$ according to (6.1), so that the normalization condition $x_R + x_H + x_C = 1$ is consistent with this evolution.

(i) When the individually optimal behavior is to be corrupt, that is, $u_C = 0, u_H = 1$, then system (6.1) written in terms of (x_H, x_C) becomes

$$\begin{cases} \dot{x}_H = (1 - x_H - x_C)r - \lambda x_H - q_{inf} x_H x_C, \\ \dot{x}_C = -x_C(b + q_{soc} x_H) + \lambda x_H + q_{inf} x_H x_C. \end{cases} \qquad (6.31)$$

Written in terms of $y = x_H - x_H^*, z = x_C - x_C^*$, it takes the form

$$\begin{cases} \dot{y} = -y(r + \lambda + q_{inf}x_C^*) - z(r + q_{inf}x_H^*) - q_{inf}yz, \\ \dot{z} = y[\lambda + (q_{inf} - q_{soc})x_C^*] + z[x_H^*(q_{inf} - q_{soc}) - b]z + (q_{inf} - q_{soc})yz. \end{cases}$$
$$(6.32)$$

The well-known condition of stability by linear approximation (Hartman theorem) states that if both eigenvalues of the linear approximation around the fixed point have real negative parts, the point is stable, and if at least one of the eigenvalues has a positive real part, the point is unstable. The requirement that both eigenvalues have negative real parts is equivalent to the requirement that the trace of the linear approximation be negative and that the determinant be positive:

$$x_H^*(q_{inf} - q_{soc}) - b - r - \lambda - q_{inf}x_C^* < 0,$$
$$\lambda(r + q_{soc}x_H^* + b) - rx_H^*(q_{inf} - q_{soc}) + br + x_C^*[r(q_{inf} - q_{soc}) + q_{inf}b] > 0$$
$$(6.33)$$

(note that the quadratic terms in x_C, x_H cancel in the second inequality). By (6.12), this can be rewritten in terms of x_H^* as

$$[x_H^*(q_{inf} - q_{soc}) - b - r - \lambda](r + b + q_{soc}x_H^*) - q_{inf}r(1 - x_H^*) < 0,$$
$$[\lambda(r + q_{soc}x_H^* + b) - rx_H^*(q_{inf} - q_{soc}) + br](r + b + q_{soc}x_H^*)$$
$$+ r(1 - x_H^*)[r(q_{inf} - q_{soc}) + q_{inf}b] > 0,$$

or in a more concise form as

$$(x_H^*)^2(q_{inf} - q_{soc})q_{soc} + x_H^*[(q_{inf} - q_{soc})(2r + b) - q_{soc}(b + \lambda)]$$
$$- (r + b)(r + b + \lambda) - rq_{inf} < 0,$$
$$(x_H^*)^2 q_{soc}[(q_{inf} - q_{soc})r - \lambda q_{soc}] + 2x_H^*(r + b)[r(q_{inf} - q_{soc})r - \lambda q_{soc}]$$
$$- r^2(q_{inf} - q_{soc}) - rbq_{inf} - (r + b)(br + r\lambda + b\lambda) < 0.$$
$$(6.34)$$

Let

$$0 \leq q_{soc} - q_{inf} \leq \frac{rq_{inf} + (r + b)(br + r\lambda + b\lambda)}{r^2}.$$

Then it is seen directly that both inequalities in (6.34) hold trivially for every positive x_H.

Assume now that

$$0 < r(q_{inf} - q_{soc}) \leq \lambda q_{soc}.$$

Then the second condition in (6.34) again holds trivially for every positive x_H. Moreover, it follows from $Q(x_H^*) = 0$ that

$$x_H^* \leq \frac{rb}{r(q_{inf} - q_{soc}) + \lambda r + \lambda b + rb} \leq \tilde{x} = \frac{b}{q_{inf} - q_{soc}}.$$

Now the left-hand side of the first inequality of (6.34) evaluated at $\tilde{x}$ is negative, because it equals

$$-\frac{bq_{soc}\lambda}{q_{inf} - q_{soc}} - r^2 - \lambda(r + b) - rq_{inf},$$

and it is also negative when evaluated at $x_H^* \le \tilde{x}$.

(ii) When individually optimal behavior is to be honest, that is, $u_C = 1, u_H = 0$, system (6.1) written in terms of (x_H, x_C) becomes

$$\begin{aligned}
\dot{x}_H &= (1 - x_H - x_C)r + \lambda x_C - q_{inf}x_H x_C, \\
\dot{x}_C &= -x_C(b + q_{soc}x_H) - \lambda x_C + q_{inf}x_H x_C.
\end{aligned} \tag{6.35}$$

To analyze the stability of the fixed point $x_H = 1, x_C = 0$, we write it in terms of x_C and $y = 1 - x_H$ as

$$\begin{aligned}
\dot{y} &= -ry + x_C(r - \lambda + q_{inf}) - q_{inf}yx_C, \\
\dot{x}_C &= x_C(q_{inf} - q_{soc} - \lambda - b) - yx_C(q_{inf} - q_{soc}).
\end{aligned}$$

According to the linear approximation, the fixed point $y = 0, x_C = 0$ of this system is stable if $q_{inf} - q_{soc} - \lambda - b < 0$, proving the first statement in (ii).

Assume that (6.14) holds. To analyze the stability of the fixed point x_H^{**}, we write system (6.35) in terms of the variables

$$y = x_H - x_H^{**} = x_H - \frac{b + \lambda}{q_{inf} - q_{soc}}, \quad z = x_C - x_C^{**} = x_C - \frac{r(q_{inf} - q_{soc} - b - \lambda)}{(r + b)q_{inf} + (\lambda - r)q_{soc}},$$

which is

$$\begin{aligned}
\dot{y} &= -y\frac{r[(r + q_{inf})(q_{inf} - q_{soc}) + \lambda q_{soc}]}{(r + b)q_{inf} + (\lambda - r)q_{soc}} - z\frac{(r + b)q_{inf} + (\lambda - r)q_{soc}}{q_{inf} - q_{soc}} - q_{inf}yz, \\
\dot{z} &= y\frac{r(q_{inf} - q_{soc} - b - \lambda)(q_{inf} - q_{soc})}{(r + b)q_{inf} + (\lambda - r)q_{soc}} + yz.
\end{aligned}$$

The characteristic equation of the matrix of linear approximation is seen to be

$$\xi^2 + \frac{r[(r + q_{inf})(q_{inf} - q_{soc}) + \lambda q_{soc}]}{(r + b)q_{inf} + (\lambda - r)q_{soc}}\xi + r(q_{inf} - q_{soc} - b - \lambda) = 0.$$

Under (6.14), both the free term and the coefficient at ξ are positive. Hence both roots have negative real parts, implying stability.   □

The results above show clearly how the presence of interaction (including social norms) influences the spread of corruption. When $q_{inf} = q_{soc} = 0$, one has one equilibrium that corresponds to corrupt or honest behavior depending on a certain relation

(6.29) between the parameters of the game. If social norms or "epidemic" myopic behavior is allowed in the model, which is quite natural for a realistic process, the situation becomes much more complicated. In particular, in a certain range of parameters, one has two stable equilibria, one corresponding to an optimally honest and another to an optimally corrupt behavior. This means in particular that similar strategies of a principal (defined by the choice of parameters b, f, w_H) can lead to quite different outcomes depending on the initial distributions of honest and corrupt agents or even on the small random fluctuations in the process of evolution. The phase transition from one to three equilibria is governed by the parameter $\bar{x}$ from (6.11). Topologically, it is equivalent to the phase transition in the famous VdW gas model.

The coefficients b and f enter exogenously in our system and can be used as tools for shifting the (precalculated) stable equilibria in the desired direction. These coefficients are not chosen strategically, which is an appropriate assumption for situations in which the principal may have only poor information about the overall distribution of states of the agents. It is of course natural to extend the model by treating the principal as a strategic optimizer who chooses b (or even can choose f) in each state to optimize a certain payoff. This would place the model in the group of MFG models with a major player, which is actively studied in the current literature.

Chapter 7
Four-State Model of Cybersecurity

Here we introduce yet another concrete MFG model in the framework of cybersecurity. It is a four-state model, and it models the response of computer owners to various offers of defense systems against a cyberhacker (for instance, a botnet attack). The model takes into account both the random process of the propagation of the infection (controlled by the botnet herder) and the decision-making process of customers. Its stationary version is again exactly solvable (but not at all trivial), though under an additional asymptotic assumption that the execution time of the decisions of the customers (say, switch on or out the defense system) is much faster that the infection rates. In particular, the phase transitions and the bifurcation points changing the number of solutions can be found explicitly.

As the calculations become harder than in the previous three-state model, we shall omit the proofs of the main results (which are more technical in nature, rather than rich in new ideas), referring for detail to the original paper [143].

7.1 The Model

Assume that any computer can be in one of four states: DI, DS, UI, US, where the first letter, D or U, refers to the state of a defended (by some system, whose effectiveness we are trying to analyze) or an unprotected computer, and the second letter, S or I, to a susceptible or infected state. The change between D and U is subject to the decisions of computer owners (though the precise time of the execution of the owner's intent is noisy), and the changes between S and I are random, with distributions depending on the level of effort v_H of the herder and the state D or U of the computer.

Let $n_{DI}, n_{DS}, n_{UI}, n_{US}$ denote the numbers of computers in the corresponding states with $N = n_{DS} + n_{DI} + n_{UI} + n_{US}$ the total number of computers. By a state of the system we shall mean either the 4-vector $n = (n_{DI}, n_{DS}, n_{UI}, n_{US})$ or its normalized version $x = (x_{DI}, x_{DS}, x_{UI}, x_{US}) = n/N$. The fraction of defended computers

133

V. N. Kolokoltsov and O. A. Malafeyev, *Many Agent Games in Socio-economic Systems: Corruption, Inspection, Coalition Building, Network Growth, Security*, Springer Series in Operations Research and Financial Engineering, https://doi.org/10.1007/978-3-030-12371-0_7

$x_{DI} + x_{DS}$ represents the analogue of the control parameter v_D from [41], the level of defense of the system, though here it results as a compound effect of individual decisions of all players.

The control parameter u of each player may have two values, 0 and 1, meaning that the player is happy with the level of defense (D or I) or prefers to switch from one to the other. When the updating decision 1 is made, the updating effectively occurs after some exponential time with the parameter λ (measuring the speed of the response of the defense system). The limit $\lambda \to \infty$ corresponds to immediate execution.

The recovery rates (the rates of change from I to S) are given constants q_{rec}^D and q_{rec}^U for defended and unprotected computers respectively, and the rates of infection from the direct attacks are $v_H q_{inf}^D$ and $v_H q_{inf}^U$ respectively with constants q_{inf}^D and q_{inf}^U. The rates of infection spreading from infected to susceptible computers are $\beta_{UU}/N, \beta_{UD}/N, \beta_{DU}/N, \beta_{DD}/N$, with numbers $\beta_{UU}, \beta_{UD}, \beta_{DU}, \beta_{DD}$, where the first (respectively second) letter in the index refers to the state of the infected (respectively susceptible) computer (the scaling $1/N$ is necessary to make the rates of unilateral changes and binary interactions comparable in the $N \to \infty$ limit).

Thus if all computers use the strategy $u_{DS}, u_{DI}, u_{US}, u_{UI}, u \in \{0, 1\}$ and the level of attack is v_H, the evolution of the frequencies x in the limit $N \to \infty$ can be described by the following kinetic equations of type (5.4) with $d = 4$:

$$
\begin{cases}
\dot{x}_{DI} = x_{DS} q_{inf}^D v_H - x_{DI} q_{rec}^D + x_{DI} x_{DS} \beta_{DD} + x_{UI} x_{DS} \beta_{UD} + \lambda(x_{UI} u_{UI} - x_{DI} u_{DI}), \\
\dot{x}_{DS} = -x_{DS} q_{inf}^D v_H + x_{DI} q_{rec}^D - x_{DI} x_{DS} \beta_{DD} - x_{UI} x_{DS} \beta_{UD} + \lambda(x_{US} u_{US} - x_{DS} u_{DS}), \\
\dot{x}_{UI} = x_{US} q_{inf}^U v_H - x_{UI} q_{rec}^U + x_{DI} x_{US} \beta_{DU} + x_{UI} x_{US} \beta_{UU} - \lambda(x_{UI} u_{UI} - x_{DI} u_{DI}), \\
\dot{x}_{US} = -x_{US} q_{inf}^U v_H + x_{UI} q_{rec}^U - x_{DI} x_{US} \beta_{DU} - x_{UI} x_{US} \beta_{UU} - \lambda(x_{US} u_{US} - x_{DS} u_{DS}).
\end{cases}
$$

$$(7.1)$$

Remark 24. *If all $\beta_{UD}, \beta_{UU}, \beta_{DU}, \beta_{UU}$ are equal to some β, $q_{inf}^D = q_{inf}^U = q^{inf}$ and $q_{rec}^D = q_{rec}^D = v_D$, where v_D is interpreted as the defender group's combined defense effort, then summing the first and the third equations in (7.1) leads to the equation*

$$\dot{x} = q_{inf} v_H (1 - x) + \beta x (1 - x) - v_D x, \qquad (7.2)$$

for the total fraction of infected computers $x = x_{DI} + x_{UI}$. This equation coincides (up to some constants) with equation (2) from [41], which is the starting point of the analysis of the paper [41].

The generator (5.8) of the corresponding Markov evolution on the states n can be written as

$$L_N F(n_{DI}, n_{DS}, n_{UI}, n_{US}) = n_{DS} q_{inf}^D v_H F(n_{DS} - 1, n_{DI} + 1)$$
$$+ n_{US} q_{inf}^U v_H F(n_{US} - 1, n_{UI} + 1)$$
$$+ n_{DI} q_{rec}^D F(n_{DI} - 1, n_{DS} + 1)$$
$$+ n_{UI} q_{rec}^U F(n_{UI} - 1, n_{US} + 1)$$
$$+ n_{DI} n_{DS} \beta_{DD} F(n_{DS} - 1, n_{DI} + 1)/N$$
$$+ n_{DI} n_{US} \beta_{DU} F(n_{US} - 1, n_{UI} + 1)/N$$
$$+ n_{UI} n_{DS} \beta_{UD} F(n_{DS} - 1, n_{DI} + 1)/N$$
$$+ n_{UI} n_{US} \beta_{UU} F(n_{US} - 1, n_{UI} + 1)/N$$
$$+ \lambda n_{DS} u_{DS} F(n_{DS} - 1, n_{US} + 1)$$
$$+ \lambda n_{US} u_{US} F(n_{US} - 1, n_{DS} + 1)$$
$$+ \lambda n_{DI} u_{DI} F(n_{DI} - 1, n_{UI} + 1)$$
$$+ \lambda n_{UI} u_{UI} F(n_{UI} - 1, n_{DI} + 1)$$

(where the unch anged values in the arguments of F on the r.h.s. are omitted), or in terms of x as

$$L_N F(x_{DI}, x_{DS}, x_{UI}, x_{US}) = N x_{DS} q_{inf}^D v_H F(x - e_{DS}/N + e_{DI}/N)$$
$$+ N x_{US} q_{inf}^U v_H F((x - e_{US}/N + e_{UI}/N)$$
$$+ N x_{DI} q_{rec}^D F(x - e_{DI}/N + e_{DS}/N)$$
$$+ N x_{UI} q_{rec}^U F(x - e_{UI}/N + e_{US}/N)$$
$$+ N x_{DI} x_{DS} \beta_{DD} F(x - e_{DS}/N + e_{DI}/N)$$
$$+ N x_{DI} x_{US} \beta_{DU} F(x - e_{US}/N + e_{UI}/N)$$
$$+ N x_{UI} x_{DS} \beta_{UD} F(x - e_{DS}/N + e_{DI}/N)$$
$$+ N x_{UI} x_{US} \beta_{UU} F(x - e_{US}/N + e_{UI}/N)$$
$$+ N \lambda x_{DS} u_{DS} F(x - e_{DS}/N + e_{US}/N)$$
$$+ N \lambda x_{US} u_{US} F(x - e_{US}/N + e_{DS}/N)$$
$$+ N \lambda x_{DI} u_{DI} F(x - e_{DI}/N + e_{UI}/N)$$
$$+ N \lambda x_{UI} u_{UI} F(x - e_{UI}/N + e_{DI}/N), \qquad (7.3)$$

where $\{e_j\}$ is the standard basis in $\mathbf{R}^4$.

If F is a differentiable function, then the generator L_N converges to the generator

$$LF(x_{DI}, x_{DS}, x_{UI}, x_{US}) = x_{DS} q_{inf}^D v_H \left(\frac{\partial F}{\partial x_{DI}} - \frac{\partial F}{\partial x_{DS}} \right) + x_{US} q_{inf}^U v_H \left(\frac{\partial F}{\partial x_{UI}} - \frac{\partial F}{\partial x_{US}} \right)$$
$$+ x_{DI} q_{rec}^D \left(\frac{\partial F}{\partial x_{DS}} - \frac{\partial F}{\partial x_{DI}} \right) + x_{UI} q_{rec}^U \left(\frac{\partial F}{\partial x_{US}} - \frac{\partial F}{\partial x_{UI}} \right)$$
$$+ x_{DI} x_{DS} \beta_{DD} \left(\frac{\partial F}{\partial x_{DI}} - \frac{\partial F}{\partial x_{DS}} \right) + x_{DI} x_{US} \beta_{DU} \left(\frac{\partial F}{\partial x_{UI}} - \frac{\partial F}{\partial x_{US}} \right)$$

$$+ x_{UI} x_{DS} \beta_{UD} \left(\frac{\partial F}{\partial x_{DI}} - \frac{\partial F}{\partial x_{DS}} \right) + x_{UI} x_{US} \beta_{UU} \left(\frac{\partial F}{\partial x_{UI}} - \frac{\partial F}{\partial x_{US}} \right)$$

$$+ \lambda x_{DS} u_{DS} \left(\frac{\partial F}{\partial x_{US}} - \frac{\partial F}{\partial x_{DS}} \right) + \lambda x_{US} u_{US} \left(\frac{\partial F}{\partial x_{DS}} - \frac{\partial F}{\partial x_{US}} \right)$$

$$+ \lambda x_{DI} u_{DI} \left(\frac{\partial F}{\partial x_{UI}} - \frac{\partial F}{\partial x_{DI}} \right) + \lambda x_{UI} u_{UI} \left(\frac{\partial F}{\partial x_{DI}} - \frac{\partial F}{\partial x_{UI}} \right) \quad (7.4)$$

in the limit $N \to \infty$. This is a first-order partial differential operator. Its characteristics are given precisely by the ODE (7.1).

If $x(t)$ and $v_H(t)$ are given, then the dynamics of each individual player is the Markov chain on four states with generator

$$L^{ind} g(DI) = \lambda u^{ind}(DI)(g(UI) - g(DI)) + q^D_{rec}(g(DS) - g(DI)),$$

$$L^{ind} g(DS) = \lambda u^{ind}(DS)(g(US) - g(DS)) + q^D_{inf} v_H(g(DI) - g(DS))$$

$$+ x_{DI} \beta_{DD}(g(DI) - g(DS)) + x_{UI} \beta_{UD}(g(DI) - g(DS)),$$

$$L^{ind} g(UI) = \lambda u^{ind}(UI)(g(DI) - g(UI)) + q^U_{rec}(g(US) - g(UI)), \quad (7.5)$$

$$L^{ind} g(US) = \lambda u^{ind}(US)(g(DS) - g(US)) + q^U_{inf} v_H(g(UI) - g(US))$$

$$+ x_{DI} \beta_{DU}(g(UI) - g(US)) + x_{UI} \beta_{UU}(g(UI) - g(US))$$

depending on the individual control u^{ind}.

Assuming that an individual pays a fee k_D per unit of time for the defense system and k_I per unit time for losses resulting from being infected, the cost during a period of time T, which is to be minimized, is

$$\int_0^T (k_D \mathbf{1}_D + k_I \mathbf{1}_I) \, ds, \quad (7.6)$$

where $\mathbf{1}_D$ (respectively $\mathbf{1}_I$) is the indicator function of the states DI, DS (respectively of the states DI, UI). Assuming that the herder has to pay $k_H v_H$ per unit of time using effort v_H and receive the income $f(x)$ depending on the distribution x of the states of the computers, the payoff, which the herder tries to maximize, is

$$\int_0^T (f_H(x) - k_H v_H) \, ds. \quad (7.7)$$

Therefore, starting with some control

$$u^{com} = (u^{com}_t(DI), u^{com}_t(DS), u^{com}_t(UI), u^{com}_t(US)),$$

the herder can find its optimal strategy $v_H(t)$ by solving the deterministic optimal control problem with dynamics (7.1) and payoff (7.7), finding both optimal v_H and the trajectory $x(t)$. Once $x(t)$ and $v_H(t)$ are known, each individual should solve the

Markov control problem (7.5) with costs (7.6), thus finding the individual optimal strategy

$$u_t^{ind} = (u_t^{ind}(DI), u_t^{ind}(DS), u_t^{ind}(UI), u_t^{ind}(US)).$$

The basic *MFG consistency equation* of Section 5.2 can be written as

$$u_t^{ind} = u_t^{com}.$$

As in the previous three-state model, instead of analyzing this complicated dynamic problem, we shall look for a simpler problem of consistent stationary strategies.

As was already pointed out in the previous chapter, there are two standard stationary problems naturally linked with a dynamic one, one being the search for the average payoff

$$g = \lim_{T \to \infty} \frac{1}{T} \int_0^T (k_D \mathbf{1}_D + k_I \mathbf{1}_I) \, dt$$

for a long-period game, and the other the search for a discounted optimal payoff. The first is governed by the solutions of HJB of the form $(T - t)\mu + g$, linear in t (then μ describing the optimal average payoff), so that g satisfies the stationary HJB equation

$$\begin{cases} \lambda \min_u u(g(UI) - g(DI)) + q_{rec}^D(g(DS) - g(DI)) + k_I + k_D = \mu, \\[4pt] \lambda \min_u u(g(US) - g(DS)) + q_{inf}^D v_H(g(DI) - g(DS)) \\[4pt] \qquad + x_{DI}\beta_{DD}(g(DI) - g(DS)) + x_{UI}\beta_{UD}(g(DI) - g(DS)) + k_D = \mu, \\[4pt] \lambda \min_u u(g(DI) - g(UI)) + q_{rec}^U(g(US) - g(UI)) + k_I = \mu, \\[4pt] \lambda \min_u u(g(DS) - g(US)) + q_{inf}^U v_H(g(UI) - g(US)) \\[4pt] \qquad + x_{DI}\beta_{DU}(g(UI) - g(US)) + x_{UI}\beta_{UU}(g(UI) - g(US)) = \mu, \end{cases} \tag{7.8}$$

where min is over two values $\{0, 1\}$. We shall denote by $u = (u_{DI}, u_{UI}, u_{DS}, u_{US})$ the argmax in this solution.

The discounted optimal payoff (with the discounting coefficient δ) satisfies the stationary HJB

$$\begin{cases} \lambda \min_u u(g(UI) - g(DI)) + q_{rec}^D(g(DS) - g(DI)) + k_I + k_D = \delta g(DI), \\[4pt] \lambda \min_u u(g(US) - g(DS)) + q_{inf}^D v_H(g(DI) - g(DS)) \\[4pt] \qquad + x_{DI}\beta_{DD}(g(DI) - g(DS)) + x_{UI}\beta_{UD}(g(DI) - g(DS)) + k_D = \delta g(DS), \\[4pt] \lambda \min_u u(g(DI) - g(UI)) + q_{rec}^U(g(US) - g(UI)) + k_I = \delta g(UI), \\[4pt] \lambda \min_u u(g(DS) - g(US)) + q_{inf}^U v_H(g(UI) - g(US)) \\[4pt] \qquad + x_{DI}\beta_{DU}(g(UI) - g(US)) + x_{UI}\beta_{UU}(g(UI) - g(US)) = \delta g(US). \end{cases} \tag{7.9}$$

For definiteness, we shall concentrate on the first one, as for the model of the previous chapter. Introducing the coefficients

$$\alpha = q_{inf}^D v_H + x_{DI}\beta_{DD} + x_{UI}\beta_{UD},$$
$$\beta = q_{inf}^U v_H + x_{DI}\beta_{DU} + x_{UI}\beta_{UU}, \tag{7.10}$$

the stationary HJB equation (7.8) can be rewritten as

$$\begin{cases} \lambda \min(g(UI) - g(DI), 0) + q_{rec}^D(g(DS) - g(DI)) + k_I + k_D = \mu, \\ \lambda \min(g(US) - g(DS), 0) + \alpha(g(DI) - g(DS)) + k_D = \mu, \\ \lambda \min(g(DI) - g(UI), 0) + q_{rec}^U(g(US) - g(UI)) + k_I = \mu, \\ \lambda \min(g(DS) - g(US), 0) + \beta(g(UI) - g(US)) = \mu, \end{cases} \tag{7.11}$$

where the choice of the first term as the minimum in these equations corresponds to the choice of control $u = 1$ (the decision to switch the strategy is made).

The *stationary MFG consistency* problem is to find $x = (x_{DI}, x_{DS}, x_{UI}, x_{US})$ and $u = (u_{DI}, u_{DS}, u_{UI}, u_{US})$, where x is the stationary point of evolution (7.1), that is,

$$\begin{cases} x_{DS}\alpha - x_{DI}q_{rec}^D + \lambda(x_{UI}u_{UI} - x_{DI}u_{DI}) = 0, \\ -x_{DS}\alpha + x_{DI}q_{rec}^D + \lambda(x_{US}u_{US} - x_{DS}u_{DS}) = 0, \\ x_{US}\beta - x_{UI}q_{rec}^U - \lambda(x_{UI}u_{UI} - x_{DI}u_{DI}) = 0, \\ -x_{US}\beta + x_{UI}q_{rec}^U - \lambda(x_{US}u_{US} - x_{DS}u_{DS}) = 0, \end{cases} \tag{7.12}$$

with $u = (u_{DI}, u_{DS}, u_{UI}, u_{US})$ giving a minimum in the solution to (7.8) or (7.11). Thus x is a fixed point of the limiting dynamics of the distribution of a large number of agents such that the corresponding stationary control is individually optimal subject to this distribution. In other words, $x = (x_{DI}, x_{DS}, x_{UI}, x_{US})$ and $u = (u_{DI}, u_{DS}, u_{UI}, u_{US})$ solve (7.11), (7.12) simultaneously.

Fixed points can practically model a stationary behavior only if they are stable. Thus we are interested in *stable solutions* (x, u) to the stationary MFG consistency problem (7.12), (7.8), where a solution is stable if the corresponding stationary distribution x is a stable equilibrium to (7.1) (with u fixed by this solution).

Apart from stability, the fixed points can be classified via their efficiency. Namely, let us say that a solution to the stationary MFG is *efficient* (or globally optimal) if the corresponding average cost μ is minimal among all other solutions.

Talking about strategies, let us reduce the discussion to nondegenerate situations, where the minima in (7.11) are achieved at a single value of u only. In principle, there are 16 possible pure stationary strategies (functions from the state space to $\{0, 1\}$). But not all of them can be realized as solutions to (7.11). In fact, if $u_{DI} = 1$, then $g(UI) < g(DI)$ (can be equal in degenerate cases), and thus $u_{UI} = 0$. This argument forbids all but four strategies as possible solutions to (7.11), namely

$$\begin{cases} (i) & g(UI) \le g(DI), \quad g(US) \le g(DS) \Longleftrightarrow u_{UI} = u_{US} = 0, \quad u_{DI} = u_{DS} = 1, \\ (ii) & g(UI) \ge g(DI), \quad g(US) \ge g(DS) \Longleftrightarrow u_{DI} = u_{DS} = 0, \quad u_{UI} = u_{US} = 1, \\ (iii) & g(UI) \le g(DI), \quad g(US) \ge g(DS) \Longleftrightarrow u_{UI} = u_{DS} = 0, \quad u_{DI} = u_{US} = 1, \\ (iv) & g(UI) \ge g(DI), \quad g(US) \le g(DS) \Longleftrightarrow u_{DI} = u_{US} = 0, \quad u_{UI} = u_{DS} = 1. \end{cases} \tag{7.13}$$

The first two strategies, either always choose U or always choose D, are acyclic, that is, the corresponding Markov processes are acyclic in the sense that there does not exist a cycle in a motion subject to these strategies. The other two strategies choose between U and D differently depending on whether a computer is infected or not. Of course, by allowing degenerate strategies, more possibilities arise.

To complete the model, let us observe that the natural assumptions on the parameters of the model arising directly from their interpretation are as follows:

$$\begin{cases} q_{rec}^D \ge q_{rec}^U, \quad q_{inf}^D < q_{inf}^U, \\ \beta_{UD} \le \beta_{UU}, \quad \beta_{DD} \le \beta_{DU}, \\ k_D \le k_I. \end{cases} \tag{7.14}$$

We shall always assume that (7.14) hold. Two additional natural simplifying assumptions that we shall use sometimes are the following: the infection rate does not depend on the level of defense of the computer transferring the infection, but only on the level of defense of the susceptible computer, that is, instead of four coefficients β one has only two of them,

$$\beta_U = \beta_{DU} = \beta_{UU}, \quad \beta_D = \beta_{UD} = \beta_{DD}, \tag{7.15}$$

and the recovery rate does not depend on whether a computer is protected against infection:

$$q_{rec} = q_{rec}^D = q_{rec}^U. \tag{7.16}$$

As we shall see, a convenient assumption, which is weaker than (7.16), turns out to be

$$q_{rec}^D - q_{rec}^U < (q_{inf}^U - q_{inf}^D)v_H. \tag{7.17}$$

Finally, it is reasonable to assume that customers can switch rather quickly their regime of defense (once they are willing to), meaning that we are effectively interested in the asymptotic regime of large λ. As we shall show, in this regime the stationary MFG problem above can be completely solved analytically. In this sense, the present model is more complicated than the model of corruption of the previous chapter, in which a transparent analytic classification of stable solutions is available already for arbitrary finite λ.

7.2 Stationary HJB Equation and Fixed Points of the Kinetic Systems

As was mentioned, the (mostly computational) proofs of the results in this section can be found in [143] and will not be reproduced here.

Two domains of x turn out to classify the solutions to the HJB equation (7.11):

$$D_1 = \{x : \beta + q_{rec}^U > \alpha + q_{rec}^D\}, \quad D_2 = \{x : \beta + q_{rec}^U < \alpha + q_{rec}^D\}.$$

More explicitly,

$$D_1 = \{x : x_{DI}(\beta_{DU} - \beta_{DD}) + x_{UI}(\beta_{UU} - \beta_{UD}) > (q_{inf}^D - q_{inf}^U)v_H + q_{rec}^D - q_{rec}^U\}.$$

By (7.14) it is seen that under a natural additional simplifying assumption, (7.16) or even (7.17), all positive x belong to D_1 (or its boundary), so that D_2 is empty.

Under the additional assumption (7.15), the condition $x \in D_1$ gets the simpler form

$$x > \bar{x} = \frac{(q_{inf}^D - q_{inf}^U)v_H + q_{rec}^D - q_{rec}^U}{\beta_U - \beta_D}. \tag{7.18}$$

To link with the conditions for cases (i)–(iv), one observes the following equivalent forms of the main condition for being in D_1:

$$\beta + q_{rec}^U > \alpha + q_{rec}^D \iff (\beta + q_{rec}^U)(\alpha + q_{rec}^D + \lambda) > (\alpha + q_{rec}^D)(\beta + q_{rec}^U + \lambda)$$

$$\iff \beta(\lambda + q_{rec}^D) - \alpha(\lambda + q_{rec}^U) > (\beta + \lambda)q_{rec}^D - (\alpha + \lambda)q_{rec}^U. \tag{7.19}$$

Setting $\varkappa = k_D/k_i$, we can summarize the properties of the HJB equation (7.11) as follows (uniqueness is always understood up to the shifts in g).

Proposition 7.2.1. *Suppose* $x \in D_1$.
(1) If

$$\frac{(\beta + \lambda)q_{rec}^D - (\alpha + \lambda)q_{rec}^U}{(\beta + q_{rec}^U + \lambda)(\alpha + q_{rec}^D)} < \varkappa < \frac{\beta(\lambda + q_{rec}^D) - \alpha(\lambda + q_{rec}^U)}{(\beta + q_{rec}^U)(\alpha + q_{rec}^D + \lambda)}, \tag{7.20}$$

then there exists a unique solution to (7.11) belonging to case (iii), and there are no other solutions to (7.11).
(2) If

$$\frac{\beta(\lambda + q_{rec}^D) - \alpha(\lambda + q_{rec}^U)}{(\beta + q_{rec}^U + \lambda)(\alpha + q_{rec}^D)} < \varkappa < \frac{(\beta + \lambda)q_{rec}^D - (\alpha + \lambda)q_{rec}^U}{(\beta + q_{rec}^U)(\alpha + q_{rec}^D + \lambda)}, \tag{7.21}$$

then there exists a unique solution to (7.11) belonging to case (iv), and there are no other solutions to (7.11).

(3) A solution belonging to case (i) exists if and only if

$$\varkappa \geq \frac{\beta(\lambda + q_{rec}^D) - \alpha(\lambda + q_{rec}^U)}{(\beta + q_{rec}^U)(\alpha + q_{rec}^D + \lambda)}, \tag{7.22}$$

and is unique if this holds. A solution belonging to case (ii) exists if and only if

$$\varkappa \leq \frac{(\beta + \lambda)q_{rec}^D - (\alpha + \lambda)q_{rec}^U}{(\beta + q_{rec}^U + \lambda)(\alpha + q_{rec}^D)}, \tag{7.23}$$

and is unique if this holds. Either condition (7.22) or (7.23) is incompatible with either (7.20) or (7.21). In particular, equation (7.11) can have at most two solutions (if both (7.22) and (7.23) hold).
(4) Under (7.16), one has always

$$\frac{\beta(\lambda + q_{rec}^D) - \alpha(\lambda + q_{rec}^U)}{(\beta + q_{rec}^U + \lambda)(\alpha + q_{rec}^D)} \geq \frac{(\beta + \lambda)q_{rec}^D - (\alpha + \lambda)q_{rec}^U}{(\beta + q_{rec}^U)(\alpha + q_{rec}^D + \lambda)}, \tag{7.24}$$

and

$$\frac{(\beta + \lambda)q_{rec}^D - (\alpha + \lambda)q_{rec}^U}{(\beta + q_{rec}^U + \lambda)(\alpha + q_{rec}^D)} \leq \frac{\beta(\lambda + q_{rec}^D) - \alpha(\lambda + q_{rec}^U)}{(\beta + q_{rec}^U)(\alpha + q_{rec}^D + \lambda)}. \tag{7.25}$$

Hence (7.21) becomes impossible, and conditions (7.22) and (7.23) become incompatible, implying the uniqueness of the solution to (7.11) for all $x \in D_1$. This unique solution belongs to cases (ii), (iii), and (i) respectively for $\varkappa$ satisfying (7.23), (7.20), (7.22) (when equality holds in (7.22) or (7.23), two solutions from different cases coincide).

Remark 25. *(1) When (7.16) does not hold, one can find situations in which solutions from cases (i) and (ii) exist simultaneously. To get simple examples, one can assume $\varkappa = 1$. (2) When two solutions exist simultaneously, one can discriminate them by the values of the average payoff μ. The values of μ arising from cases (i) and (ii) are usually different. (3) The uniqueness result under (7.16) is quite remarkable, since it does not seem to follow a priori from any intuitive arguments.*

Proposition 7.2.2. *Suppose $x \in D_2$.*
(1) If

$$\varkappa > \frac{(\beta + \lambda)q_{rec}^D - (\alpha + \lambda)q_{rec}^U}{(\beta + q_{rec}^U)(\alpha + q_{rec}^D + \lambda)}, \tag{7.26}$$

then there exists a unique solution to (7.11) belonging to case (i), and there are no other solutions to (7.11).
(2) If

$$\varkappa < \frac{\beta(\lambda + q_{rec}^D) - \alpha(\lambda + q_{rec}^U)}{(\beta + q_{rec}^U + \lambda)(\alpha + q_{rec}^D)}, \tag{7.27}$$

then there exists a unique solution to (7.11) belonging to case (ii), and there are no
other solutions to (7.11).

(3) A solution belonging to case (iii) exists if and only if

$$\frac{(\beta + \lambda)q_{rec}^D - (\alpha + \lambda)q_{rec}^U}{(\beta + q_{rec}^U + \lambda)(\alpha + q_{rec}^D)} \leq \varkappa \leq \frac{\beta(\lambda + q_{rec}^D) - \alpha(\lambda + q_{rec}^U)}{(\beta + q_{rec}^U)(\alpha + q_{rec}^D + \lambda)}, \tag{7.28}$$

and is unique if this holds. A solution belonging to case (iv) exists if and only if

$$\frac{\beta(\lambda + q_{rec}^D) - \alpha(\lambda + q_{rec}^U)}{(\beta + q_{rec}^U + \lambda)(\alpha + q_{rec}^D)} \leq \varkappa \leq \frac{(\beta + \lambda)q_{rec}^D - (\alpha + \lambda)q_{rec}^U}{(\beta + q_{rec}^U)(\alpha + q_{rec}^D + \lambda)}, \tag{7.29}$$

and is unique if this holds. Each of the conditions (7.28) and (7.29) is incompatible
with either (7.26) or (7.27). In particular, equation (7.11) may have at most two
solutions (if (7.28) and (7.29) hold simultaneously).

Essential simplifications that allow eventually for a full classification of the stationary MFG consistency problem occur in the limit of large λ. For a precise formulation in case

$$\delta = q_{rec}^D - q_{rec}^U > 0, \tag{7.30}$$

one needs further decomposition of the domains D_1, D_2. Namely, for $j = 1, 2$, let

$$D_{j1} = \{x \in D_j : \frac{\delta}{\alpha + q_{rec}^D} < \frac{\beta - \alpha}{\beta + q_{rec}^U}\}.$$

Proposition 7.2.3. *The following hold for large λ outside an interval of $\varkappa$ of size of
order λ^{-1}:*

*(1) Under (7.16), conditions (7.23), (7.20), (7.22) classifying the solutions to the
HJB equation can be rewritten as*

$$\varkappa \leq 0, \quad 0 < \varkappa < \frac{(\beta - \alpha)}{\beta + q_{rec}^U}, \quad \varkappa \geq \frac{(\beta - \alpha)}{\beta + q_{rec}^U}, \tag{7.31}$$

respectively. In particular, solutions of case (ii) become impossible.

*(2) Suppose $x \in D_1$ and (7.30) holds. If $x \in D_{11}$, then there exists a unique solu-
tion to (7.11), which belongs to cases (ii), (iii), (i) for*

$$\varkappa < \frac{\delta}{\alpha + q_{rec}^D}, \quad \frac{\delta}{\alpha + q_{rec}^D} < \varkappa < \frac{\beta - \alpha}{\beta + q_{rec}^U}, \quad \varkappa > \frac{\beta - \alpha}{\beta + q_{rec}^U}, \tag{7.32}$$

*respectively. If $x \in D_{12}$, then solutions from case (iii) do not exist, and there exist
two solutions to (7.11) for*

$$\frac{\beta - \alpha}{\beta + q_{rec}^U} < \varkappa < \frac{\delta}{\alpha + q_{rec}^D}, \tag{7.33}$$

belonging to cases (i) and (ii), and only one solution otherwise.

(3) Suppose $x \in D_2$. If $x \in D_{22}$, then solutions from case (iii) do not exist, and there is always a unique solution to (7.11) belonging to case (ii), (iv), or (i), for

$$\varkappa < \frac{\beta - \alpha}{\alpha + q^D_{rec}}, \quad \frac{\beta - \alpha}{\alpha + q^D_{rec}} < \varkappa < \frac{\delta}{\beta + q^U_{rec}}, \quad \varkappa > \frac{\delta}{\beta + q^U_{rec}}, \tag{7.34}$$

respectively. If $x \in D_{21}$, then there are two solutions to (7.11) for

$$\frac{\delta}{\alpha + q^D_{rec}} < \varkappa < \frac{\beta - \alpha}{\beta + q^U_{rec}}, \tag{7.35}$$

which belong to cases (iii) and (iv), and one solution otherwise. This unique solution belongs to case (ii) or (i) for

$$\varkappa < \frac{\beta - \alpha}{\alpha + q^D_{rec}}, \quad \varkappa > \frac{\delta}{\beta + q^U_{rec}}$$

respectively and to case (iv) otherwise.

Next let us solve the fixed-point system (7.12).

Proposition 7.2.4. *(1) There exists a unique solution to system (7.12) with the strategy U being individually optimal (that is, with the first acyclic stationary strategy $u_{UI} = u_{US} = 0$, $u_{DI} = u_{DS} = 1$), and it is stable. It equals $x = (0, 0, x^*_{UI}, 1 - x^*_{UI})$, with x^*_{UI} given by*

$$x^* = x^*_{UI} = \frac{1}{2\beta_{UU}} \Big[\beta_{UU} - q^U_{rec} - q^U_{inf} v_H$$
$$+ \sqrt{(\beta_{UU} + q^U_{inf} v_H)^2 + (q^U_{rec})^2 - 2q^U_{rec}(\beta_{UU} - q^U_{inf} v_H)} \Big]. \tag{7.36}$$

(2) There exists a unique solution to system (7.12) with the strategy D being individually optimal (that is, with the second acyclic stationary strategy), and it is stable. It equals $x = (x^*_{DI}, 1 - x^*_{DI}, 0, 0)$, with x^*_{DI} being the unique solution of equation

$$Q_D(y) = \beta_{DD} y^2 + y(q^D_{rec} - \beta_{DD} + q^D_{inf} v_H) - q^D_{inf} v_H = 0 \tag{7.37}$$

on the interval $(0, 1)$, that is,

$$x^*_{DI} = \frac{1}{2\beta_{DD}} \Big[\beta_{DD} - q^D_{rec} - q^D_{inf} v_H$$
$$+ \sqrt{(\beta_{DD} + q^D_{inf} v_H)^2 + (q^D_{rec})^2 - 2q^D_{rec}(\beta_{DD} - q^D_{inf} v_H)} \Big].$$

For the case of cyclic strategies, the asymptotic regime $\lambda \to \infty$ will be employed.

Proposition 7.2.5. *(1) For large λ, there exists a unique solution to the system (7.12) in case (iii), that is, with $u_{UI} = u_{DS} = 0$, $u_{DI} = u_{US} = 1$, and it is stable. It has the form $x = (0, 1 - \bar{x}^*_{UI}, \bar{x}^*_{UI}, 0)$ up to corrections of order λ^{-1}, with $\bar{x}^*_{UI}$ being the unique solution of*

$$\beta_{UD} x^2_{UI} + x_{UI}(q^U_{rec} - \beta_{UD} + q^D_{inf} v_H) - q^D_{inf} v_H = O(\lambda^{-1}) \tag{7.38}$$

on $(0, 1)$ given by

$$\bar{x}^*_{UI} = O(\lambda^{-1})$$

$$+ \frac{1}{2\beta_{UD}} \left[\beta_{UD} - q^U_{rec} - q^D_{inf} v_H + \sqrt{(\beta_{UD} + q^D_{inf} v_H)^2 + (q^U_{rec})^2 - 2q^U_{rec}(\beta_{UD} - q^D_{inf} v_H)} \right]. \tag{7.39}$$

(2) For large λ, there exists a unique solution to system (7.12) in case (iv), that is, with $u_{UI} = u_{DS} = 1$, $u_{DI} = u_{US} = 0$, and it is stable. It has the form $x = (\bar{x}^*_{DI}, 0, 0, 1 - \bar{x}^*_{UI})$ up to corrections of order λ^{-1}, with $\bar{x}^*_{DI}$ being the unique solution of

$$\beta_{DU} x^2_{DI} + x_{DI}(q^U_{rec} - \beta_{DU} + q^U_{inf} v_H) - q^U_{inf} v_H = O(\lambda^{-1}), \tag{7.40}$$

on $(0, 1)$.

7.3 Solutions to the Stationary MFG Problem

Combining Propositions 7.2.4, 7.2.5, and 7.2.3 allows one to fully characterize the solutions to our stationary MFG consistency problem for large λ.

The most straightforward general conclusion is the following.

Theorem 7.3.1. *For large λ, there may exist up to four solutions to the stationary MFG problem, with only one in each of the cases (i)–(iv). All these solutions are stable.*

Remark 26. *Notice that this statement is not at all obvious a priori, and may not be true for finite λ, where solutions to case (iii) or (iv) are found from an equation of fourth order.*

As an example of a more precise classification, let us present the stationary MFG problem under assumption (7.17), which ensures that all solutions lie in the domain D_1.

Let us introduce the function

$$\varkappa(z) = \frac{(q^U_{inf} - q^D_{inf})v_H + z(\beta_{UU} - \beta_{UD})}{q^U_{inf}v_H + z\beta_{UU} + q^U_{rec}}.$$

First, let (7.16) hold. It is seen from Propositions 7.2.4 and 7.2.5 that for large λ (and apart from $\varkappa$ from negligible intervals of size of order λ^{-1} that we shall ignore), a solution of the stationary MFG problem exists in case (i) if

$$\varkappa > \varkappa^* = \varkappa(x^*_{UI}),$$

and a solution of the stationary MFG problem exists in case (iii) if

$$\varkappa < \bar{\varkappa}^* = \varkappa(\bar{x}^*_{UI}),$$

where x^*_{UI} and $\bar{x}^*_{UI}$ are given by (7.36) and (7.39) respectively. Thus one can have up to two (automatically stable) solutions to the stationary MFG problem. Let us make this number precise.

Differentiating $\varkappa(z)$, we find directly that it is increasing if

$$\beta_{UU}(q^D_{inf}v_H + q^U_{rec}) > \beta_{UD}(q^U_{inf}v_H + q^U_{rec}), \tag{7.41}$$

and decreasing otherwise. Hence the relation $\varkappa^* > \bar{\varkappa}^*$ is equivalent to the same or the opposite relation for x^*_{UI} and $\bar{x}^*_{UI}$. Thus we are led to the following conclusion.

Theorem 7.3.2. *Let (7.16) hold.*

(1) If (7.41) holds and $x^*_{UI} > \bar{x}^*_{UI}$, or the opposite to (7.41) holds and $x^*_{UI} < \bar{x}^*_{UI}$, then $\varkappa^* > \bar{\varkappa}^*$. Consequently, for $\varkappa < \bar{\varkappa}^*$, there exists a unique solution to the stationary MFG problem, which is stable and belongs to case (iii); for $\varkappa \in (\bar{\varkappa}^*, \varkappa^*)$, there are no solutions to the stationary MFG problem; for $\varkappa > \varkappa^*$, there exists a unique solution to the stationary MFG problem, which is stable and belongs to case (i).

(2) If (7.41) holds and $x^*_{UI} < \bar{x}^*_{UI}$, or the opposite to (7.41) holds and $x^*_{UI} > \bar{x}^*_{UI}$, then $\varkappa^* < \bar{\varkappa}^*$. Consequently, for $\varkappa < \varkappa^*$, there exists a unique solution to the stationary MFG problem, which is stable and belongs to case (iii); for $\varkappa \in (\varkappa^*, \bar{\varkappa}^*)$, there exist two (stable) solutions to the stationary MFG problem; for $\varkappa > \bar{\varkappa}^*$, there exists a unique solution to the stationary MFG problem, which is stable and belongs to case (i).

Thus if one considers the system for all parameters fixed except for $\varkappa$ (essentially specifying the price of the defense service), then the points $\varkappa^*$ and $\bar{\varkappa}^*$ are the bifurcation points, where the phase transitions occur.

To deal with the case in which (7.16) does not hold, let us introduce the numbers

$$\varkappa_1 = \frac{\beta - \alpha}{\beta + q^U_{rec}}(x^*_{UI}), \quad \varkappa_2 = \frac{\delta}{\alpha + q^D_{rec}}(x^*_{DI}), \quad \varkappa_3 = \frac{\delta}{\alpha + q^D_{rec}}(\bar{x}^*_{UI}), \quad \varkappa_4 = \frac{\beta - \alpha}{\beta + q^U_{rec}}(\bar{x}^*_{UI}),$$

where $x_{UI}^*, x_{DI}^*, \bar{x}_{UI}^*$ in parentheses mean that α, β defined in (7.10) are evaluated at the corresponding solutions given by Propositions 7.2.4 and 7.2.5. Since $x_{UI}^*, x_{DI}^*, \bar{x}_{UI}^*$ are expressed in terms of different parameters, any order relation between them are to be expected in general. Of course, restrictions appear under additional assumptions, for instance $x_{DI}^* = \bar{x}_{UI}^*$ under (7.15). From Proposition 7.2.3 we deduce the following.

Theorem 7.3.3. *Let (7.17) and (7.30) hold.*

Depending on the order relation between $x_{UI}^*, x_{DI}^*, \bar{x}_{UI}^*$, one can have up to three solutions to the stationary MFG problem for large λ, the characterization in each case being fully explicit, since for $\varkappa > \varkappa_1$, there exists a unique solution in case (i), for $\varkappa < \varkappa_2$, there exists a unique solution in case (ii), for $\varkappa_3 < \varkappa < \varkappa_4$, there exists a unique solution in case (iii).

Thus in this case the points $\varkappa_1, \varkappa_2, \varkappa_3, \varkappa_4$ are the bifurcation points, where the phase transitions occur.

The situation in which (7.17) does not hold is analogous, though there appears an additional bifurcation relating to x crossing the border between D_1 and D_2, and the possibility of having four solutions arises.

Chapter 8
Turnpikes for MFGs on Two-Dimensional Networks

Here we develop a more general mean-field-game model for two-dimensional arrays of states, extending the models of Chapters 6 and 7 and providing full details for the simplest case of $2d$ states, $d \in \mathbf{N}$. In order to tackle new technical difficulties arising from a larger state space, we introduce new asymptotic regimes, small discount, and small interaction asymptotics.

In our two-dimensional arrays of states, one of the dimensions will be controlled mostly by the decision of agents (say, the level of tax evasion in the context of inspection games) and the other by a principal (major player) or evolutionary interactions (say, the level of agents in a bureaucratic ladder, the type of computer virus used by a botnet herder).

We shall dwell upon two basic interpretations of our model: corrupt bureaucrats playing against the principal (say, governmental representative, also referred to in the literature as a benevolent dictator) or computer owners playing against a botnet herder (which then takes the role of the principal), which tries to infect the computers with viruses. Other interpretations can be done, for instance in the framework of inspection games (inspector and taxpayers) or the spread of a disease in epidemiology (among animals or humans), or the defense against a biological weapon, or other models of Chapter 1.

In this chapter we essentially strengthen the results on stability. In contrast to previous chapters, we are not dealing just with the dynamic stability of the stationary points of kinetic equations (with fixed control), but with the full stability of the backward–forward system. This analysis allows us to build a wide class of time-dependent solutions to the corresponding nonstationary MFG such that the stationary solution plays the role of a turnpike for this class.

© Springer Nature Switzerland AG 2019
V. N. Kolokoltsov and O. A. Malafeyev, *Many Agent Games in Socio-economic Systems: Corruption, Inspection, Coalition Building, Network Growth, Security*, Springer Series in Operations Research and Financial Engineering, https://doi.org/10.1007/978-3-030-12371-0_8

8.1   The Setting for $2d$-State Models

We assume that every agent in the group of N agents has $2d$ states: iI and iS, where $i \in \{1, \cdots, d\}$ is referred to as a strategy. In the first interpretation, the letters S and I designate the senior and initial positions of a bureaucrat in the hierarchical ladder, and i designates the level or type of corrupt behavior (say, the level of bribes one asks from customers or, more generally, the level of illegal profit sought). In the literature on corruption, the state I is often denoted by R and is referred to as the reserved state. It is interpreted as a job of the lowest salary given to untrustworthy bureaucrats. In the second interpretation, the letters S and I designate susceptible and infected states of computers, and i denotes the level or the type of defense system available on the market.

We assume that the choice of a strategy depends exclusively on the decision of an agent. The control parameter u of each player may have d values denoting the strategy the agent prefers at a given moment. As long as this coincides with the current strategy, the updating of a strategy does not occur. Once the decision to change i to j is made, the actual updating is supposed to occur at a certain rate λ. As in the previous chapter, we shall be mostly interested in the asymptotic regime of fast execution of individual decisions, that is, $\lambda \to \infty$.

The change between S and I may have two causes: the action of the principal (pressure game component) and that of the peers (evolutionary component). In the first interpretation, the principal can promote the bureaucrats from the initial to the senior position or degrade them to the reserved initial position whenever their illegal behavior is discovered. The peers can also take part in this process, contributing to the degrading of corrupt bureaucrats, for instance when they trespass against certain social norms. In the second interpretation, the principal, the botnet herder, infects computers with a virus by direct attacks, turning S to I, and the virus then spreads through the network of computers by a pairwise interaction. The recovery change from I to S is due to some system of repairs, which can be different at different protection levels i.

Let q_+^i denote the recovery rates of upgrading from iI to iS, and q_-^i the rates of degradation (punishment or infection) from state iS to iI, which are independent of the states of other agents (pressure component), and let β_{ij}/N denote the rates at which any agent in state iI can stimulate the degradation (punishment or infection) of another agent from jS to jI (evolutionary component). For simplicity we ignore here the possibility of upgrading changes from jS to jI due to the interaction with peers.

In the detailed description of our model, its states are N-tuples $\{\omega_1, \cdots, \omega_N\}$, with each ω_j being either iS or jI, describing the positions of all N players of the game. If each player $l \in \{1, \cdots, N\}$ has a strategy u_l, the system evolves according to a continuous-time Markov chain, with the transitions occurring at the rates specified above. When the fees w_I^i and w_S^i for staying in the corresponding states per unit time are specified together with the terminal payments $g_T(iI)$, $g_T(iS)$ at each state at the terminal time T, we are in the setting of a stochastic dynamic game of N players.

In a symmetric Nash equilibrium, all players are supposed to play the same strategy. In this case, players become indistinguishable, leading to a reduced description of the game, in which the state space is the set $\mathbf{Z}_+^{2d}$ of vectors

$$n = (n_{\{iI\}}, n_{\{iS\}}) = (n_{1I}, \cdots, n_{dI}, n_{1S}, \cdots, n_{dS})$$

with coordinates representing the number of agents in the corresponding states. Alternatively, in the normalized version, the state space becomes the subset of the standard simplex Σ_{2d}^N in $\mathbf{R}^{2d}$ consisting of the vectors

$$x = (x_I, x_S) = (x_{1I}, \cdots, x_{dI}, x_{1S}, \cdots, x_{dS}) = n/N,$$

with $N = n_{1S} + n_{1I} + \cdots + n_{dS} + n_{dI}$ the total number of agents. The functions f on $\mathbf{Z}_+^{2d}$ and F on Σ_{2d}^N are supposed to be linked by the scaling transformation $f(n) = F(n/N)$.

Assuming that all players have the same strategy $u_t^{com} = \{u_t^{com}(iS), u_t^{com}(iI)\}$, the Markov chain introduced above reduces to a time-inhomogeneous Markov chain on $\mathbf{Z}_+^{2d}$ or Σ_{2d}^N, which can be described by its (time-dependent) generator. Omitting the unchanged values in the arguments of F on the r.h.s., this generator can be written as

$$L_N^t f(n) = \lambda n_{jI} \sum_{j,i} \mathbf{1}(u_t^{com}(jI) = i)[f(n_{jI} - 1, n_{iI} + 1) - f(n)]$$

$$+ \lambda n_{jI} \sum_{j,i} \mathbf{1}(u_t^{com}(jS) = i)[f(n_{js} - 1, n_{iS} + 1) - f(n)]$$

$$+ \sum_j n_{jI} q_j^+ [f(n_{jI} - 1, n_{js} + 1) - f(n)] + \sum_j n_{js} q_j^- [f(n_{js} - 1, n_{jI} + 1) - f(n)]$$

$$+ \frac{1}{N} \sum_{j,i} n_{iI} n_{js} \beta_{ij} [f(n_{jI} - 1 n_{js} + 1) - f(n)],$$

for the Markov chain on $\mathbf{Z}_+^{2d}$, and as

$$L_N^t F(x) = \lambda N x_{jI} \sum_{j,i} \mathbf{1}(u_t^{com}(jI) = i)[F(x - e_{jI}/N + e_{iI}/N) - F(x)]$$

$$+ \lambda N x_{jI} \sum_{j,i} \mathbf{1}(u_t^{com}(jS) = i)[F(x - e_{js}/N + e_{iS}/N) - F(x)]$$

$$+ N \sum_j x_{jI} q_j^+ [F(x - e_{jI}/N + e_{js}/N) - F(x)] + N \sum_j x_{js} q_j^- [F(x - e_{js}/N + e_{jI}/N) - F(x)]$$

$$+ N \sum_{j,i} x_{iI} x_{js} \beta_{ij} [F(x - e_{jI}/N + e_{js}/N) - F(x)]$$

for the Markov chain on Σ_{2d}^{N}. Here and below, $\mathbf{1}(M)$ denotes the indicator function of a set M, and $\{e_{jI}, e_{iS}\}$ is the standard orthonormal basis in $\mathbf{R}^{2d}$.

With this idea in mind, assuming that F is differentiable, and expanding it in a Taylor series, one finds that as $N \to \infty$, the last generators tend to

$$
L^t F(x) = \lambda x_{jI} \sum_{j,i} \mathbf{1}(u_t^{com}(jI) = i) \left[\frac{\partial F}{\partial x_{iI}} - \frac{\partial F}{\partial x_{jI}} \right] + \lambda x_{jI} \sum_{j,i} \mathbf{1}(u_t^{com}(jS) = i) \left[\frac{\partial F}{\partial x_{iS}} - \frac{\partial F}{\partial x_{jS}} \right]
$$

$$
+ \sum_{j} x_{jI} q_j^+ \left[\frac{\partial F}{\partial x_{jS}} - \frac{\partial F}{\partial x_{jI}} \right] + \sum_{j} x_{jS} q_j^- \left[\frac{\partial F}{\partial x_{jI}} - \frac{\partial F}{\partial x_{jS}} \right] + \sum_{j,i} x_{iI} x_{jS} \beta_{ij} \left[\frac{\partial F}{\partial x_{jI}} - \frac{\partial F}{\partial x_{jS}} \right].
$$

This operator L^t is a first-order partial differential operator, which therefore generates a deterministic Markov process, whose dynamics is given by the system of characteristics arising from L^t:

$$
\dot{x}_{iI} = \lambda \sum_{j \neq i} x_{jI} \mathbf{1}(u^{com}(jI) = i) - \lambda \sum_{j \neq i} x_{iI} \mathbf{1}(u^{com}(iI) = j) + x_{iS} q_-^i - x_{iI} q_+^i + \sum_{j} x_{iS} x_{jI} \beta_{ji},
$$

$$
\dot{x}_{iS} = \lambda \sum_{j \neq i} x_{jS} \mathbf{1}(u^{com}(jS) = i) - \lambda \sum_{j \neq i} x_{iS} \mathbf{1}(u^{com}(iS) = j) - x_{iS} q_-^i + x_{iI} q_+^i - \sum_{j} x_{iS} x_{jI} \beta_{ji},
$$

$$
\tag{8.1}
$$

for all $i = 1, \cdots, d$ (of course, all x_{iS}, x_{iI} are functions of time t).

The optimal behavior of agents depends on the payoffs in different states, the terminal payoff, and possibly the costs of transitions. For simplicity, we shall ignore here the latter. In talking about corrupt agents, it is natural to talk about maximizing profit, while in talking about infected computers, it is natural to talk about minimizing costs. To unify the exposition, we shall deal with the minimization of costs, which is equivalent to the maximization of their opposite values.

Recall that w_I^i and w_S^i denote the costs per time unit of staying in iI and iS respectively. According to our interpretation of S as a better state, $w_S^i < w_I^i$ for all i.

Given the evolution of the states $x = x(s)$ of the whole system in a time interval $[t, T]$, the individually optimal costs $g(iI)$ and $g(iS)$ and individually optimal controls $u_s^{ind}(iI)$ and $u_s^{ind}(iS)$ of an arbitrary agent can be found from the Bellman or HJB equation of type (5.3):

$$
\dot{g}_t(iI) + \lambda \min_u \sum_{j=1}^{d} \mathbf{1}(u(iI) = j)(g_t(jI) - g_t(iI)) + q_+^i(g_t(iS) - g_t(iI)) + w_I^i = 0,
$$

$$
\dot{g}_t(iS) + \lambda \min_u \sum_{j=1}^{d} \mathbf{1}(u(iS) = j)(g_t(jS) - g_t(iS)) + q_-^i(g_t(iI) - g_t(iS))
$$

$$
+ \sum_{j=1}^{d} \beta_{ji} x_{jI}(s)(g_t(iI) - g_t(iS)) + w_S^i = 0,
$$

$$
\tag{8.2}
$$

which holds for all i and is complemented by the terminal condition $g_T(iI)$, $g_T(iS)$.

The basic *MFG consistency equation* of Section 5.2 for a time interval $[t, T]$ can now be written as $u_s^{com} = u_s^{ind}$.

In this chapter we shall mostly work with a discounted payoff with the discounting coefficient $\delta > 0$, in which case the HJB equation for the discounted optimal payoff $e^{-t\delta} g_t$ of an individual player with any time horizon T can be written as (by putting $e^{-t\delta} g_t$ instead of g_t in (8.2))

$$\dot{g}_t(iI) + \lambda \min_u \sum_{j=1}^d \mathbf{1}(u(iI) = j)(g_t(jI) - g_t(iI)) + q_+^i(g_t(iS) - g_t(iI)) + w_I^i = \delta g_t(iI),$$

$$\dot{g}_t(iS) + \lambda \min_u \sum_{j=1}^d \mathbf{1}(u(iS) = j)(g_t(jS) - g_t(iS)) + q_-^i(g_t(iI) - g_t(iS))$$

$$+ \sum_{j=1}^d \beta_{ji} x_{jI}(s)(g_t(iI) - g_t(iS)) + w_S^i = \delta g_t(iS),$$

$$(8.3)$$

which holds for all i and is complemented by certain terminal conditions $g_T(iI)$, $g_T(iS)$.

For the discounted payoff, the basic *MFG consistency equation* $u_s^{com} = u_s^{ind}$ for a time interval $[t, T]$ can be reformulated by saying that x, u, g solve the coupled forward–backward system (8.1), (8.3), so that the u_s^{com} used in (8.1) coincide with the minimizers in (8.3). The main objective of this chapter is to provide a general class of solutions of the discounted MFG consistency equation with stationary (time-independent) controls u^{com}.

As a first step to this objective we shall analyze the fully stationary solutions, whereby the evolution (8.1) is replaced by the corresponding fixed-point condition:

$$\lambda \sum_{j \neq i} x_{jI} \mathbf{1}(u^{com}(jI) = i) - \lambda \sum_{j \neq i} x_{iI} \mathbf{1}(u^{com}(iI) = j) + x_{iS} q_-^i - x_{iI} q_+^i + \sum_j x_{iS} x_{jI} \beta_{ji} = 0,$$

$$\lambda \sum_{j \neq i} x_{jS} \mathbf{1}(u^{com}(jS) = i) - \lambda \sum_{j \neq i} x_{iS} \mathbf{1}(u^{com}(iS) = j) - x_{iS} q_-^i + x_{iI} q_+^i - \sum_j x_{iS} x_{jI} \beta_{ji} = 0.$$

$$(8.4)$$

In contrast to our treatment of the models of the previous chapters, we shall concentrate on the stationary solutions arising from the discounted payoff, which is simpler to link with the time-dependent solutions. If the discounting coefficient is δ, then the stationary discounted optimal payoff g'satisfies the stationary version of (8.3):

$$\lambda \min_u \sum_{j=1}^d \mathbf{1}(u(iI) = j)(g(jI) - g(iI)) + q_+^i(g(iS) - g(iI)) + w_I^i = \delta g(iI),$$

$$\lambda \min_u \sum_{j=1}^d \mathbf{1}(u(iS) = j)(g(jS) - g(iS)) + q_-^i(g(iI) - g(iS))$$

$$+ \sum_{j=1}^{d} \beta_{ji} x_{jI} (g(iI) - g(iS)) + w_S^i = \delta g(iS). \tag{8.5}$$

The *stationary MFG consistency condition* is the coupled system of equations (8.4) and (8.5), so that the individually optimal stationary control u^{ind} found from (8.5) coincides with the common stationary control u^{com} from (8.4).

For simplicity, we shall be interested in *nondegenerate controls u^{ind}* characterized by the condition that the minimum in (8.5) is always attained on a single value of u.

Remark 27. *(i) The nondegeneracy assumption is common in MFG modeling, since the argmax of the control provides the coupling with the forward equation on the states, where nonuniqueness creates a nontrivial question of choosing a representative. (ii) In our case, nondegeneracy is a very natural assumption of the "general position." It is equivalent to the assumption that for a solution g, the minimum of $g(iI)$ is achieved on only one i, and the minimum of $g(jS)$ is also achieved on only one j. Any "strange" coincidence of two values of g can be removed by arbitrary weak changes in the parameters of the model. (iii) If nondegeneracy is relaxed, we get of course much more complicated solutions, with the support of the equilibrium distributed between the states providing the minima of $g(iI)$ and $g(iS)$.*

A new technical novelty as compared with the models of Chapters 6 and 7 will be systematic working in the asymptotic regimes of small discount δ and small interaction coefficients β_{ij}. This approach leads to more or less explicit calculations of stationary MFG solutions and their further justification.

8.2 Stationary MFG for $2d$-State Models

We start by identifying all possible stationary nondegenerate controls that can occur as solutions of (8.5). Let $[i(I), k(S)]$ denote the following strategy: switch to strategy i when in I and to k when in S, that is, $u(jI) = i$ and $u(jS) = k$ for all j.

Proposition 8.2.1. *Nondegenerate controls solving* (8.5) *can be only of type* $[i(I), k(S)]$.

Proof. Let i be the unique minimum point of $g(iI)$, and k the unique minimum point of $g(iS)$. Then the optimal strategy is $[i(I), k(S)]$. □

Let us consider first the control $[i(I), i(S)]$, denoting it by $\hat{u}^i$:

$$\hat{u}^i(jS) = \hat{u}^i(jI) = i, \quad j = 1, \cdots, d.$$

We shall refer to the control $\hat{u}^i$ as the one with the strategy i individually optimal.

The control $\hat{u}^i$ and the corresponding distribution x solve the stationary MFG problem if they solve the corresponding HJB (8.5), that is,

$$\begin{cases} q^i_+(g(iS) - g(iI)) + w^i_I = \delta g(iI), \\ q^i_-(g(iI) - g(iS)) + \sum_k \beta_{ki} x_{kI}(g(iI) - g(iS)) + w^i_S = \delta g(iS), \\ \lambda(g(iI) - g(jI)) + q^j_+(g(jS) - g(jI)) + w^j_I = \delta g(jI), \quad j \neq i, \\ \lambda(g(iS) - g(jS)) + q^j_-(g(jI) - g(jS)) \\ \qquad + \sum_k \beta_{kj} x_{kI}(g(jI) - g(jS)) + w^j_S = \delta g(jS), \quad j \neq i, \end{cases} \tag{8.6}$$

where for all $j \neq i$,

$$g(iI) \leq g(jI), \quad g(iS) \leq g(jS), \tag{8.7}$$

and x is a fixed point of the evolution (8.4) with $u^{com} = \hat{u}^i$, that is,

$$\begin{cases} x_{iS} q^i_- - x_{iI} q^i_+ + \sum_j x_{iS} x_{jI} \beta_{ji} + \lambda \sum_{j \neq i} x_{jI} = 0, \\ -x_{iS} q^i_- + x_{iI} q^i_+ - \sum_j x_{iS} x_{jI} \beta_{ji} + \lambda \sum_{j \neq i} x_{jS} = 0, \\ x_{jS} q^j_- - x_{jI} q^j_+ + \sum_k x_{jS} x_{kI} \beta_{kj} - \lambda x_{jI} = 0, \quad j \neq i, \\ -x_{jS} q^j_- + x_{jI} q^j_+ - \sum_k x_{jS} x_{kI} \beta_{kj} - \lambda x_{jS} = 0, \quad j \neq i. \end{cases} \tag{8.8}$$

This solution $(\hat{u}^i, x)$ is stable if x is a stable fixed point of the evolution (8.1) with $u^{com} = \hat{u}^i$, that is, of the evolution

$$\begin{cases} \dot{x}_{iI} = x_{iS} q^i_- - x_{iI} q^i_+ + \sum_j x_{iS} x_{jI} \beta_{ji} + \lambda \sum_{j \neq i} x_{jI}, \\ \dot{x}_{iS} = -x_{iS} q^i_- + x_{iI} q^i_+ - \sum_j x_{iS} x_{jI} \beta_{ji} + \lambda \sum_{j \neq i} x_{jS}, \\ \dot{x}_{jI} = x_{jS} q^j_- - x_{jI} q^j_+ + \sum_k x_{jS} x_{kI} \beta_{kj} - \lambda x_{jI}, \quad j \neq i, \\ \dot{x}_{jS} = -x_{jS} q^j_- + x_{jI} q^j_+ - \sum_k x_{jS} x_{kI} \beta_{kj} - \lambda x_{jS}, \quad j \neq i. \end{cases} \tag{8.9}$$

Adding together the last two equations of (8.8), we find that $x_{jI} = x_{jS} = 0$ for $j \neq i$, as one would expect. Consequently, the whole system (8.8) reduces to the single equation

$$x_{iS} q^i_- + x_{iI} \beta_{ii} x_{iS} - x_{iI} q^i_+ = 0,$$

which for $y = x_{iI}$, $1 - y = x_{iS}$, yields the quadratic equation

$$Q(y) = \beta_{ii} y^2 + y(q_+^i - \beta_{ii} + q_-^i) - q_-^i = 0,$$

with the unique solution on the interval $(0, 1)$

$$x^* = \frac{1}{2\beta_{ii}} \left[\beta_{ii} - q_+^i - q_-^i + \sqrt{(\beta_{ii} + q_-^i)^2 + (q_+^i)^2 - 2q_+^i(\beta_{ii} - q_-^i)} \right]. \quad (8.10)$$

To analyze stability of the fixed point $x_{iI} = x^*$, $x_{iS} = 1 - x^*$ and $x_{jI} = x_{jS} = 0$ for $j \neq i$, we introduce the variables $y = x_{iI} - x^*$. In terms of y and x_{jI}, x_{jS} with $j \neq i$, the system (8.9) can be rewritten as

$$\begin{cases} \dot{y} = [1 - x^* - y - \sum_{j \neq i}(x_{jI} + x_{jS})][q_- + \sum_{k \neq i} x_{kI}\beta_{ki} + (y + x^*)\beta_{ii}] - (y + x^*)q_+^i + \lambda \sum_{j \neq i} x_{jI}, \\ \dot{x}_{jI} = x_{jS}[q_-^j + \sum_{k \neq i} x_{kI}\beta_{kj} + (y + x^*)\beta_{ij}] - x_{jI}q_+^j - \lambda x_{jI}, \quad j \neq i, \\ \dot{x}_{jS} = -x_{jS}[q_-^j + \sum_{k \neq i} x_{kI}\beta_{kj} + (y + x^*)\beta_{ij}] + x_{jI}q_+^j - \lambda x_{jS}, \quad j \neq i. \end{cases}$$

$$(8.11)$$

Its linearized version around the fixed point zero is

$$\begin{cases} \dot{y} = (1 - x^*)(\sum_{k \neq i} x_{kI}\beta_{ki} + y\beta_{ii}) - [y + \sum_{k \neq i}(x_{kI} + x_{kS})](q_-^i + x^*\beta_{ii}) - yq_+^i + \sum_{k \neq i} \lambda x_{kI}, \\ \dot{x}_{jI} = x_{jS}(q_-^j + x^*\beta_{ij}) - x_{jI}q_+^j - \lambda x_{jI}, \quad j \neq i, \\ \dot{x}_{jS} = -x_{jS}(q_-^j + x^*\beta_{ij}) + x_{jI}q_+^j - \lambda x_{jS}, \quad j \neq i. \end{cases}$$

Since the equations for x_{jI}, x_{jS} contain neither y nor other variables, the eigenvalues of this linear system are

$$\xi_i = (1 - 2x^*)\beta_{ii} - q_-^i - q_+^i,$$

and $(d - 1)$ pairs of eigenvalues arising from $(d - 1)$ systems

$$\begin{cases} \dot{x}_{jI} = x_{jS}(q_-^j + x^*\beta_{ij}) - x_{jI}q_+^j - \lambda x_{jI}, \quad j \neq i, \\ \dot{x}_{jS} = -x_{jS}(q_-^j + x^*\beta_{ij}) + x_{jI}q_+^j - \lambda x_{jS}, \quad j \neq i, \end{cases}$$

that is,

$$\begin{cases} \xi_1^j = -\lambda - (q_+^j + q_-^j + x^*\beta_{ii}) \\ \xi_2^j = -\lambda. \end{cases}$$

These eigenvalues being always negative, the condition of stability is reduced to the negativity of the first eigenvalue ξ_i:

$$2x^* > 1 - \frac{q_+^i + q_-^i}{\beta_{ii}}.$$

But this is true due to (8.10), implying that this fixed point is always stable (by the Grobman–Hartman theorem).

Next, the HJB equation (8.6) takes the form

$$\begin{cases} q_+^i(g(iS) - g(iI)) + w_I^i = \delta g(iI), \\ q_-^i(g(iI) - g(iS)) + \beta_{ii}x^*(g(iI) - g(iS)) + w_S^i = \delta g(iS), \\ \lambda(g(iI) - g(jI)) + q_+^j(g(jS) - g(jI)) + w_I^j = \delta g(jI), \quad j \neq i, \\ \lambda(g(iS) - g(jS)) + q_-^j(g(jI) - g(jS)) \\ \quad + \beta_{ij}x^*(g(jI) - g(jS)) + w_S^j = \delta g(jS), \quad j \neq i. \end{cases} \quad (8.12)$$

Subtracting the first equation from the second yields

$$g(iI) - g(iS) = \frac{w_I^i - w_S^i}{q_-^i + q_+^i + \beta_{ii}x^* + \delta}. \quad (8.13)$$

In particular, $g(iI) > g(iS)$ always, as expected. Next, by the first equation of (8.12),

$$\delta g(iI) = w_I^i - \frac{q_+^i(w_I^i - w_S^i)}{q_-^i + q_+^i + \beta_{ii}x^* + \delta}. \quad (8.14)$$

Consequently,

$$\delta g(iS) = w_I^i - \frac{(q_+^i + \delta)(w_I^i - w_S^i)}{q_-^i + q_+^i + \beta_{ii}x^* + \delta} = w_S^i + \frac{(q_-^i + \beta_{ii}x^*)(w_I^i - w_S^i)}{q_-^i + q_+^i + \beta_{ii}x^* + \delta}. \quad (8.15)$$

Subtracting the third equation of (8.12) from the fourth yields

$$(\lambda + q_+^j + q_-^j + \beta_{ii}x^* + \delta)(g(jI) - g(jS)) - \lambda(g(iI) - g(iS)) = w_I^i - w_S^i,$$

implying

$$g(jI) - g(jS) = \frac{w_I^j - w_S^j + \lambda(g(iI) - g(iS))}{\lambda + q_+^j + q_-^j + \beta_{ij}x^* + \delta}$$

$$= g(iI) - g(iS) + [(w_I^j - w_S^j) - (g(iI) - g(iS))(q_+^j + q_-^j + \beta_{ij}x^* + \delta)]\lambda^{-1} + O(\lambda^{-2}). \quad (8.16)$$

From the fourth equation of (8.12), it now follows that

$$(\delta + \lambda)g(jI) = w_I^j - q_+^j(g(jI) - g(jS)) + \lambda g(iI),$$

so that

$$g(jI) = g(iI) + [w_I^j - q_+^j(g(iI) - g(iS)) - \delta g(iI)]\lambda^{-1} + O(\lambda^{-2}). \quad (8.17)$$

Consequently,

$$g(jS) = g(jI) - (g(jI) - g(jS))$$

$$= g(iS) + [w_S^j + (q_-^j + \beta_{ii}x^*)(g(iI) - g(iS)) - \delta g(iS)]\lambda^{-1} + O(\lambda^{-2}). \quad (8.18)$$

Thus conditions (8.7) in the main order of asymptotics in $\lambda \to \infty$ become

$$w_I^j - q_+^j(g(iI) - g(iS)) - \delta g(iI) \geq 0, \quad w_S^j + (q_-^j + \beta_{ii}x^*)(g(iI) - g(iS)) - \delta g(iS) \geq 0,$$

or equivalently,

$$w_I^j - w_I^i \geq \frac{(q_+^j - q_+^i)(w_I^i - w_S^i)}{q_-^i + q_+^i + \beta_{ii}x^* + \delta}, \quad w_S^j - w_S^i \geq \frac{[q_-^i - q_-^j + (\beta_{ii} - \beta_{ij})x^*](w_I^i - w_S^i)}{q_-^i + q_+^i + \beta_{ii}x^* + \delta}. \quad (8.19)$$

To first order in small β_{ij}, this takes on the following simpler form, independent of x^*:

$$\frac{w_I^j - w_I^i}{w_I^i - w_S^i} \geq \frac{q_+^j - q_+^i}{q_-^i + q_+^i + \delta}, \quad \frac{w_S^j - w_S^i}{w_I^i - w_S^i} \geq \frac{q_-^i - q_-^j}{q_-^i + q_+^i + \delta}. \quad (8.20)$$

Conversely, if the inequalities in (8.20) are strict, then (8.19) also holds with the strict inequalities for sufficiently small β_{ij}. Consequently, (8.7) also holds with the strict inequalities.

Summarizing, we have proved the following.

Proposition 8.2.2. *If* (8.20) *holds for all* $j \neq i$ *with strict inequality, then for sufficiently large* λ *and sufficiently small* β_{ij}, *there exists a unique solution to the stationary MFG consistency problem* (8.4) *and* (8.5) *with the optimal control* $\hat{u}^i$; *the stationary distribution is* $x_i^I = x^*, x_i^S = 1 - x^*$ *with* x^* *given by* (8.10), *and it is stable; the optimal payoffs are given by* (8.14), (8.15), (8.17), (8.18). *Conversely, if for all sufficiently large* λ *there exists a solution to the stationary MFG consistency problem* (8.4) *and* (8.5) *with the optimal control* $\hat{u}^i$, *then* (8.19) *holds.*

Let us turn to the control $[i(I), k(S)]$ with $k \neq i$, denoting it by $\hat{u}^{i,k}$:

$$\hat{u}^{i,k}(jS) = k, \quad \hat{u}^{i,k}(jI) = i, \quad j = 1, \cdots, d.$$

The fixed-point condition under $u^{com} = \hat{u}^{i,k}$ takes the form

$$\begin{cases} x_{iS}q_-^i - x_{iI}q_+^i + \sum_j x_{iS}x_{jI}\beta_{ji} + \lambda \sum_{j \neq i} x_{jI} = 0 \\ - x_{iS}q_-^i + x_{iI}q_+^i - \sum_j x_{iS}x_{jI}\beta_{ji} - \lambda x_{iS} = 0 \\ x_{kS}q_-^k - x_{kI}q_+^k + \sum_j x_{kS}x_{jI}\beta_{jk} - \lambda x_{kI} = 0 \\ - x_{kS}q_-^i + x_{kI}q_+^k - \sum_j x_{kS}x_{jI}\beta_{jk} + \lambda \sum_{j \neq k} x_{jS} = 0 \\ x_{lS}q_-^l - x_{lI}q_+^l + \sum_j x_{lS}x_{jI}\beta_{jl} - \lambda x_{lS} = 0 \\ - x_{lS}q_-^l + x_{lI}q_+^l - \sum_j x_{lS}x_{jI}\beta_{jl} - \lambda x_{lI} = 0, \end{cases} \tag{8.21}$$

where $l \neq i, k$.

Adding the last two equations yields $x_{lI} + x_{lS} = 0$, and hence $x_{lI} = x_{lS} = 0$ for all $l \neq i, k$, as one would expect. Consequently, for indices i, k, the system takes the form

$$\begin{cases} x_{iS}q_-^i - x_{iI}q_+^i + x_{iS}x_{iI}\beta_{ii} + x_{iS}x_{kI}\beta_{ki} + \lambda x_{kI} = 0, \\ - x_{iS}q_-^i + x_{iI}q_+^i - x_{iS}x_{iI}\beta_{ii} - x_{iS}x_{kI}\beta_{ki} - \lambda x_{iS} = 0, \\ x_{kS}q_-^k - x_{kI}q_+^k + x_{kS}x_{kI}\beta_{kk} + x_{kS}x_{iI}\beta_{ik} - \lambda x_{kI} = 0, \\ - x_{kS}q_-^i + x_{kI}q_+^k - x_{kS}x_{kI}\beta_{kk} - x_{kS}x_{iI}\beta_{ik} + \lambda x_{iS} = 0. \end{cases} \tag{8.22}$$

Adding the first two equations (or the last two equations) yields $x_{kI} = x_{iS}$. Since by normalization

$$x_{kS} = 1 - x_{iS} - x_{kI} - x_{iI} = 1 - x_{iI} - 2x_{kI},$$

we are left with two equations only:

$$\begin{cases} x_{kI}q_-^i - x_{iI}q_+^i + x_{kI}x_{iI}\beta_{ii} + x_{kI}^2\beta_{ki} + \lambda x_{kI} = 0, \\ (1 - x_{iI} - 2x_{kI})(q_-^k + x_{kI}\beta_{kk} + x_{iI}\beta_{ik}) - (\lambda + q_+^k)x_{kI} = 0. \end{cases} \tag{8.23}$$

From the first equation, we obtain

$$x_{iI} = \frac{\lambda x_{kI} + \beta_{ki}x_{kI}^2 + q_-^i x_{kI}}{q_+^i - x_{kI}\beta_{ii}} = \frac{\lambda x_{kI}}{q_+^i - x_{kI}\beta_{ii}}(1 + O(\lambda^{-1})).$$

Hence x_{kI} is of order $1/\lambda$, and therefore

$$x_{iI} = \frac{\lambda x_{kI}}{q_+^i}(1 + O(\lambda^{-1})) \iff x_{kI} = \frac{x_{iI} q_+^i}{\lambda}(1 + O(\lambda^{-1})). \tag{8.24}$$

In the major order in large-λ asymptotics, the second equation of (8.23) yields

$$(1 - x_{iI})(q_-^k + \beta_{ik} x_{iI}) - q_+^i x_{iI} = 0,$$

or for $y = x_{iI}$,

$$Q(y) = \beta_{ik} y^2 + y(q_+^i - \beta_{ik} + q_-^k) - q_-^k = 0,$$

which is effectively the same equation as the one that appeared in the analysis of the control $[i(I), i(S)]$. It has a unique solution on the interval $(0, 1)$:

$$x_{iI}^* = \frac{1}{2\beta_{ik}}\left[\beta_{ik} - q_+^i - q_-^k + \sqrt{(\beta_{ik} + q_-^k)^2 + (q_+^i)^2 - 2q_+^i(\beta_{ik} - q_-^k)}\right]. \tag{8.25}$$

Let us note that for small β_{ik}, it can be expanded as

$$x_{iI}^* = \frac{q_-^k}{q_-^k + q_+^i} + O(\beta) = \frac{q_-^k}{q_-^k + q_+^i} + \frac{q_-^k q_+^i}{(q_-^k + q_+^i)^3}\beta + O(\beta^2). \tag{8.26}$$

Similar (a bit more lengthy) calculations to those for the control $[i(I), i(S)]$ show that the obtained fixed point of the evolution (8.1) is always stable. We omit the details.

Let us turn to the HJB equation (8.6), which under the control $[i(I), k(S)]$ takes the form

$$\begin{cases} q_+^i(g(iS) - g(iI)) + w_I^i = \delta g(iI), \\ \lambda(g(kS) - g(iS)) + \tilde{q}_-^i(g(iI) - g(iS)) + w_S^i = \delta g(iS), \\ \lambda(g(iI) - g(kI)) + q_+^k(g(kS) - g(kI)) + w_I^k = \delta g(kI), \\ \tilde{q}_-^k(g(kI) - g(kS)) + w_S^k = \delta g(kS) \\ \lambda(g(iI) - g(jI)) + q_+^j(g(jS) - g(jI)) + w_I^j = \delta g(jI), \quad j \neq i, k, \\ \lambda(g(kS) - g(jS)) + \tilde{q}_-^j(g(jI) - g(jS)) + w_S^j = \delta g(jS), \quad j \neq i, k, \end{cases} \tag{8.27}$$

supplemented by the consistency condition

$$g(iI) \leq g(jI), \quad g(kS) \leq g(jS), \tag{8.28}$$

for all j, where we have introduced the notation

$$\tilde{q}_-^j = \tilde{q}_-^j(i, k) = q_-^j + \beta_{ij} x_{iI} + \beta_{kj} x_{kI}. \tag{8.29}$$

The first four equations do not depend on the rest of the system and can be solved independently. To begin with, we use the first and fourth equations to obtain

$$g(iS) = g(iI) + \frac{\delta g(iI) - w_I^i}{q_+^i}, \quad g(kI) = g(kS) + \frac{\delta g(kS) - w_S^k}{\tilde{q}_-^k}. \tag{8.30}$$

Then the second and third equations can be written as the following system for the variables $g(kS)$ and $g(iI)$:

$$\begin{cases} \lambda g(kS) - (\lambda + \delta)g(iI) - (\lambda + \delta + \tilde{q}_-^i)\dfrac{\delta g(iI) - w_I^i}{q_+^i} + w_S^i = 0 \\[2mm] \lambda g(iI) - (\lambda + \delta)g(kS) - (\lambda + \delta + q_+^k)\dfrac{\delta g(kS) - w_S^k}{\tilde{q}_-^k} + w_I^k = 0, \end{cases}$$

or more simply as

$$\begin{cases} \lambda q_+^i g(kS) - [\lambda(q_+^i + \delta) + \delta(q_+^i + \tilde{q}_-^i + \delta)]g(iI) = -w_I^i(\lambda + \delta + \tilde{q}_-^i) - w_S^i q_+^i \\[2mm] [\lambda(\tilde{q}_-^k + \delta) + \delta(\tilde{q}_-^k + q_+^k + \delta)]g(kS) - \lambda \tilde{q}_-^k g(iI) = w_I^k \tilde{q}_-^k + w_S^k(\lambda + \delta + q_+^k). \end{cases}$$
$$\tag{8.31}$$

Let us find the asymptotic behavior of the solution for large λ. To this end, let us write

$$g(iS) = g^0(iS) + \frac{g^1(iS)}{\lambda} + O(\lambda^{-2}),$$

with similar notation for other values of g. Dividing (8.31) by λ and preserving only the leading terms in λ, we get the system

$$\begin{cases} q_+^i g^0(kS) - (q_+^i + \delta)g^0(iI) = -w_I^i, \\[2mm] (\tilde{q}_-^k + \delta)g^0(kS) - \tilde{q}_-^k g^0(iI) = w_S^k. \end{cases} \tag{8.32}$$

Solving this system and using (8.30) to find the corresponding leading terms $g^0(iS)$, $g^0(kI)$ yields

$$g^0(iS) = g^0(kS) = \frac{1}{\delta}\frac{\tilde{q}_-^k w_I^i + q_+^i w_S^k + \delta w_S^k}{\tilde{q}_-^k + q_+^i + \delta},$$

$$g^0(kI) = g^0(iI) = \frac{1}{\delta}\frac{\tilde{q}_-^k w_I^i + q_+^i w_S^k + \delta w_I^i}{\tilde{q}_-^k + q_+^i + \delta}. \tag{8.33}$$

The remarkable equations $g^0(iS) = g^0(kS)$ and $g^0(kI) = g^0(iI)$ arising from the calculations have a natural interpretation: for instantaneous execution of personal decisions, discrimination between strategies i and j is not possible. Thus to get the conditions ensuring (8.28), we have to look for the next order of expansion in λ.
Keeping in (8.31) the terms of zero order in $1/\lambda$ yields the system

$$\begin{cases} q_+^i g^1(kS) - (q_+^i + \delta)g^1(iI) = \delta(q_+^i + \tilde{q}_-^i + \delta)g^0(iI) - w_I^i(\delta + \tilde{q}_-^i) - w_S^i q_+^i \\ (\tilde{q}_-^k + \delta)g^1(kS) - \tilde{q}_-^k g^1(iI) = -\delta(\tilde{q}_-^k + q_+^k + \delta)g^0(kS) + w_I^k \tilde{q}_-^k + w_S^k(\delta + q_+^k). \end{cases}$$

$$(8.34)$$

Taking into account (8.33), the conditions $g(iI) \leq g(kI)$ and $g(kS) \leq g(iS)$ become

$$\tilde{q}_-^k g^1(iI) \leq g^1(kS)(\tilde{q}_-^k + \delta), \quad q_+^i g^1(kS) \leq g^1(iI)(q_+^i + \delta). \quad (8.35)$$

Solving (8.34), we obtain

$$g^1(kS)\delta(\tilde{q}_-^k + q_+^i + \delta) = \tilde{q}_-^k[q_+^i w_S^i + (q_+^i + \delta)w_I^k + (\tilde{q}_-^i - \tilde{q}_-^k - q_+^i - \delta)w_I^i]$$
$$+ [q_+^i(q_+^k - \tilde{q}_-^k - q_+^i) + \delta(q_+^k - q_+^i)]w_S^k,$$
$$g^1(iI)\delta(\tilde{q}_-^k + q_+^i + \delta) = q_+^i[\tilde{q}_-^k w_I^k + (\tilde{q}_-^k + \delta)w_S^i + (q_+^k - q_+^i - \tilde{q}_-^k - \delta)w_S^k]$$
$$+ [\tilde{q}_-^k(\tilde{q}_-^i - q_+^i - \tilde{q}_-^k) + \delta(\tilde{q}_-^i - \tilde{q}_-^k)]w_I^i.$$

$$(8.36)$$

We can now check the conditions (8.35). Remarkably enough, the right- and left-hand sides of both inequalities always coincide for $\delta = 0$, so that the actual condition arises from comparing higher-order terms in δ. To first order with respect to the expansion in small δ, the conditions (8.35) take the following simple form:

$$\tilde{q}_-^k(w_I^k - w_I^i) + w_S^k(q_+^k - q_+^i) \geq 0, \quad q_+^i(w_S^i - w_S^k) + w_I^i(\tilde{q}_-^i - \tilde{q}_-^k) \geq 0. \quad (8.37)$$

From the last two equations of (8.27) we can find $g(jS)$ and $g(jI)$ for $j \neq i, k$, yielding

$$g(jI) = g(iI) + \frac{1}{\lambda}[w_I^j - \delta g(iI) + q_+^j(g(iI) - g(kS))] + O(\lambda^{-2}),$$
$$g(jS) = g(kS) + \frac{1}{\lambda}[w_S^j - \delta g(kS) + \tilde{q}_-^j(g(iI) - g(kS))] + O(\lambda^{-2}).$$

$$(8.38)$$

From these equations we can derive the rest of the conditions (8.28), namely that $g(iI) \leq g(jI)$ for $j \neq k$ and $g(kS) \leq g(jS)$ for $j \neq i$. To first order in the small-δ expansion, they become

$$q_+^j(w_I^i - w_S^k) + w_I^j(\tilde{q}_-^k + q_+^i) \geq 0, \quad \tilde{q}_-^j(w_I^i - w_S^k) + w_S^j(\tilde{q}_-^k + q_+^i) \geq 0. \quad (8.39)$$

Since for small β_{ij}, the difference $\tilde{q}_-^j - q_-^j$ is small, we have proved the following result.

Proposition 8.2.3. *Assume*

$$q_+^j(w_I^i - w_S^k) + w_I^j(q_-^k + q_+^i) > 0, \quad j \neq k,$$
$$q_-^j(w_I^i - w_S^k) + w_S^j(q_-^k + q_+^i) > 0, \quad j \neq i,$$
$$q_-^k(w_I^k - w_I^i) + w_S^k(q_+^k - q_+^i) > 0, \quad q_+^i(w_S^i - w_S^k) + w_I^i(q_-^i - q_-^k) > 0.$$

$$(8.40)$$

Then for sufficiently large λ, small δ, and small β_{ij}, there exists a unique solution to the stationary MFG consistency problem (8.4) and (8.5) with the optimal control $\hat{u}^{i,k}$, the stationary distribution is concentrated on strategies i and k, with x_{iI}^* given by (8.25) or (8.26) up to terms of order $O(\lambda^{-1})$, and it is stable; the optimal payoffs are given by (8.33), (8.36), (8.38).

Conversely, if for all sufficiently large λ and small δ there exists a solution to the stationary MFG consistency problem (8.4) and (8.5) with the optimal control $\hat{u}^{i,k}$, then (8.37) and (8.39) hold.

8.3 Time-Dependent MFG Problems and Turnpikes

Solutions given by Propositions 8.2.2 and 8.2.3 work only when the initial distribution and terminal payoff are exactly those given by the stationary solution. What happens for other initial conditions? The stability results of Propositions 8.2.2 and 8.2.3 represent only a step in the right direction here, since they ensure stability only under the assumption that all (or almost all) players use from the very beginning the corresponding stationary control, which might not be the case. To analyze the stability properly, we have to consider the full time-dependent problem. For possibly time-varying evolution $x(t)$ of the distribution, the time-dependent HJB equation for the discounted optimal payoff $e^{-t\delta}g$ of an individual player with any time horizon T has the form (8.3).

In order to have a solution with a stationary u, we have to show that solving the linear equation obtained from (8.3) by fixing this control will be consistent in the sense that this control will actually give a minimum in (8.3) at all times.

For definiteness, let us concentrate on the stationary control $\hat{u}^i$, the corresponding linear equation being given the form

$$
\begin{cases}
\dot{g}(iI) + q_+^i(g(iS) - g(iI)) + w_I^i = \delta g(iI), \\
\dot{g}(iS) + q_-^i(g(iI) - g(iS)) + \sum_k \beta_{ki} x_{kI}(t)(g(iI) - g(iS)) + w_S^i = \delta g(iS), \\
\dot{g}(jI) + \lambda(g(iI) - g(jI)) + q_+^j(g(jS) - g(jI)) + w_I^j = \delta g(jI), \quad j \neq i, \\
\dot{g}(jS) + \lambda(g(iS) - g(jS)) + q_-^j(g(jI) - g(jS)) \\
\qquad + \sum_k \beta_{kj} x_{kI}(t)(g(jI) - g(jS)) + w_S^j = \delta g(jS), \quad j \neq i
\end{cases}
$$

(8.41)

(the dependence of g on t is omitted for brevity), with the supplementary requirement (8.7), but which has to hold now for a time-dependent solution g.

Theorem 8.3.1. *Assume that the strengthened form of (8.20) holds, that is,*

$$
\frac{w_I^j - w_I^i}{w_I^i - w_S^i} > \frac{q_+^j - q_+^i}{q_-^i + q_+^i + \delta}, \quad \frac{w_S^j - w_S^i}{w_I^i - w_S^i} > \frac{q_-^i - q_-^j}{q_-^i + q_+^i + \delta}
$$

(8.42)

for all $j \neq i$. Assume, moreover, that

$$w_I^j - w_I^i \geq 0, \quad w_S^j - w_S^i \geq 0, \tag{8.43}$$

for all $j \neq i$. Then for all $\lambda > 0$ and all sufficiently small β_{ij}, the following holds. For every $T > t$, initial distribution $x(t)$, and terminal values g_T such that $g_T(jI) - g_T(jS) \geq 0$ for all j, $g_T(iI) - g_T(iS)$ is sufficiently small, and

$$g_T(iI) \leq g_T(jI) \quad \text{and} \quad g_T(iS) \leq g_T(jS), \quad j \neq i, \tag{8.44}$$

there exists a unique solution to the discounted MFG consistency equation such that u is stationary and equals $\hat{u}^i$ everywhere. Moreover, this solution is such that for large $T - t$, $x(s)$ tends to the fixed point of Proposition 8.2.2 for $s \to T$ and g_s remains near the stationary solution of Proposition 8.2.2 for almost all time, apart from a small initial period around t and some final period around T.

Remark 28. *The condition (8.43) is natural: having better fees in a nonoptimal state may create instability. In fact, with the terminal g_T vanishing, it is seen from (8.45) below that if $w_I^j - w_I^i < 0$, then the solution will be directly ejected from the region $g(iI) \leq g(jI)$, so that the stability of this region would be destroyed. It is interesting to ask what kind of solutions to the forward–backward system one could construct if $w_I^j - w_I^i < 0$ occurs.*

Proof. To show that starting with the terminal condition belonging to the cone specified by (8.44) we will remain in this cone for all $t \leq T$, it is sufficient to prove that on every boundary point of this cone that can be achieved by the evolution, the inverted tangent vector of the system (8.41) is not directed outside of the cone. This result is more or less obvious and is well known (see, e.g., Chapter 3 of [142] for details). From (8.41) we find that

$$\dot{g}(jI) - \dot{g}(iI) = (\lambda + \delta)(g(jI) - g(iI)) + q_+^j(g(jI) - g(jS))$$
$$- q_+^i(g(iI) - g(iS)) - (w_I^j - w_I^i). \tag{8.45}$$

Therefore, the condition for staying inside the cone (8.44) for a boundary point with $g(jI) = g(iI)$ reads $\dot{g}(jI) - \dot{g}(iI) \leq 0$, or

$$(w_I^j - w_I^i) \geq q_+^j(g(jI) - g(jS)) - q_+^i(g(iI) - g(iS)). \tag{8.46}$$

Since $g(jI) = g(iI)$,

$$0 \leq g(jI) - g(jS) \leq g(iI) - g(iS).$$

Therefore, if $q_+^i \geq q_+^j$, the r.h.s. of (8.46) is nonpositive, and hence (8.46) holds by the first assumption of (8.43). Hence we can assume further that $q_+^i < q_+^j$.
Again by

$$0 \leq g(jI) - g(jS) \leq g(iI) - g(iS),$$

a simpler sufficient condition for (8.46) is

$$(w_I^j - w_I^i) \geq (q_+^j - q_+^i)(g(iI) - g(iS)). \tag{8.47}$$

Subtracting the first two equations of (8.41), we find that

$$\dot{g}(iI) - \dot{g}(iS) = a(s)(g(iI) - g(iS)) - (w_I^i - w_S^i)$$

with

$$a(t) = q_+^i + q_-^i + \delta + \sum_k \beta_{ki} x_{kI}(t).$$

Consequently,

$$g_t(iI) - g_t(iS) = \exp\{-\int_t^T a(s)\,ds\}(g_T(iI) - g_T(iS))$$

$$+ (w_I^i - w_S^i)\int_t^T \exp\{-\int_t^s a(\tau)\,d\tau\}ds. \tag{8.48}$$

Therefore, condition (8.47) will be fulfilled for all sufficiently small $g_T(iI) - g_T(iS)$ whenever

$$(w_I^j - w_I^i) > (q_+^j - q_+^i)(w_I^i - w_S^i)\int_t^T \exp\{-\int_t^s a(\tau)\,d\tau\}ds. \tag{8.49}$$

But since $a(t) \geq q_+^i + q_-^i + \delta$, we have

$$\exp\{-\int_t^s a(\tau)\,d\tau\} \leq \exp\{-(s-t)(q_+^i + q_-^i + \delta)\},$$

so that (8.47) holds if

$$\frac{w_I^j - w_I^i}{w_I^i - w_S^i} \geq \frac{q_+^j - q_+^i}{q_+^i + q_-^i + \delta}\left(1 - \exp\{-(T-t)(q_+^i + q_-^i + \delta)\}\right), \tag{8.50}$$

which is true under the first assumptions of (8.42), because $q_+^i < q_+^j$.

Similarly, to study a boundary point with $g(jS) = g(iS)$, we find that

$$\dot{g}(jS) - \dot{g}(iS) = (\lambda + \delta)(g(jS) - g(iS)) - (q_-^j + \sum_k \beta_{kj} x_{kI})(g(jI) - g(jS))$$

$$+ (q_-^i + \sum_k \beta_{ki} x_{kI})(g(iI) - g(iS)) - (w_S^j - w_S^i).$$

Therefore, the condition for staying inside the cone (8.44) for a boundary point with $g(jS) = g(iS)$ reads

$$(w_S^j - w_S^i) \geq (q_-^i + \sum_k \beta_{ki} x_{kI})(g(iI) - g(iS)) - (q_-^j + \sum_k \beta_{kj} x_{kI})(g(jI) - g(jS)).$$
(8.51)

Now $0 \leq g(iI) - g(iS) \leq g(jI) - g(jS)$, so that (8.51) is fulfilled if

$$(w_S^j - w_S^i) \geq (q_-^i + \sum_k \beta_{ki} x_{kI} - q_-^j - \sum_k \beta_{kj} x_{kI})(g(iI) - g(iS)) \qquad (8.52)$$

for all times. Taking into account the requirement that all β_{ij} be sufficiently small, we find as above that the condition holds under the second assumptions of (8.42) and (8.43).

Similarly (but actually even more simply), one shows that the condition $g_t(jI) - g_t(jS) \geq 0$ remains valid for all times and all j.

The last statement of the theorem concerning $x(s)$ follows from the observation that the eigenvalues of the linearized evolution $x(s)$ are negative and well separated from zero, implying the global stability of the fixed point of the evolution for sufficiently small β. The last statement of the theorem concerning $g(s)$ follows by a similar stability argument for the linear evolution (8.41), taking into account that away from the initial point t, the trajectory $x(t)$ stays arbitrary close to its fixed point. $\square$

The last property of the solution from Theorem 8.3.1 can be expressed by saying that the stationary solution provides a so-called turnpike for the time-dependent solution; see, e.g., [152] and [235] for reviews in the stochastic and deterministic settings. A similar time-dependent class of turnpike solutions can be constructed from the stationary control of Proposition 8.2.3.

8.4 Sketch of Extensions to General Two-Dimensional Networks

Here we sketch a general framework for interactions in two-dimensional networks that include binary interaction (evolutionary game component), individual decision-making inside the environment (mean-field-game component), the action of the principal or the major player (pressure and resistance game component), and noise. We shall see how the stationary problem can be solved explicitly under rather general assumptions and discuss the specific features of these solutions. We will not give details of the construction of time-dependent solutions with a turnpike property (analogous to the results of the models with $2d$ states above), which can be found in [126].

Let $H = \{1, \cdots, |H| = n\}$ be a finite set characterizing the hierarchy of agents, say the position in the bureaucratic ladder of an organization. Alternatively, it can describe the geographical (location) distribution of agents.

Let $B = \{1, \cdots, |B|\}$ be a finite set characterizing the types of behavior or a strategy of agents, say the level of compliance with official regulations, or the level of involvement in terrorist activity, or the level of protection of a computer against cyberhackers. The states of an agent are given by the pairs (h, b) with $h \in H, b \in B$.

Remark 29. *In some cases it is reasonable to include an additional zero state, kind of a sink, where no choice of B is available, say a terrorist put in jail or a corrupt agent degraded to the lowest reserved salary without any power to exercise corrupt behavior. Thus the state space can be either $S = H \times B$ or $\tilde{S} = H \times B \cup \{0\} = S \cup \{0\}$. We shall stick here with the first option.*

The *decision structure* (B, E_D, λ) is an unoriented graph with vertex set B and edge set E_D, where an edge e joins i and j whenever an agent has power to switch between i and j according to its personal decision. Such a switch requires a certain random λ-exponential time. A single parameter λ for all switching is chosen for simplicity. As we have mentioned, we shall mostly look at the regime $\lambda \to \infty$. As usual, we shall assume the agents to be homogeneous and indistinguishable, in the sense that their strategies and payoffs may depend only on their states and not on any other individual characteristics (say, their names). A decision of an agent in a state (h, b) at any time is given by the *decision matrix* $u = (u_{h,b \to \tilde{b}}), \tilde{b} \neq b$, expressing that agent's interest in switching from $b \to \tilde{b}$. In our model, we shall consider agents with clear decisions, namely, for all h, b, either the decision vector $(u_{h,b \to \tilde{b}})$ is identically zero, if the agent does not want to change the strategy b, or there exists $b_1 \neq b$ such that $u_{h,b \to b_1} = 1$ and all other coordinates of $(u_{h,b \to \tilde{b}})$ vanish, if the agent wishes to move from b to b_1.

The *pressure structure* $(H, E_P, q_{j \to i,b})$ is an oriented graph such that an edge e joins j and i if the principal has the right to transfer the agents from j to i. The coefficients $q_{j \to i,b}$ represent the rates of such transitions (the waiting time is a $q_{j \to i,b}$-exponential waiting time). Of course, these rates can depend on some control parameter of the principal, but we shall not exploit this possibility here.

The *evolution structure* describes in general the change in the distribution of states due to the pairwise interaction of agents (exchange of opinions, fights with competitors, effects of social norms), which can be described by the set of rates $q_{s_1 \to s_2}^s$, by which an agent in the state s can stimulate the transition of another agent from s_1 to s_2. For instance, an honest agent can help the principal to discover and punish the illegal behavior of a corrupt agent. The transitions $s_1 \to s_2$ can be naturally separated into transitions in B and in H, yielding *behavioral* or *hierarchical evolution structures*. The evolutionary changes in B represent an alternative to the individual-decision-based changes described by the decision structure (B, E_D, λ) and can be considered negligible in the limit $\lambda \to \infty$ that we shall look at here. Thus taking into account a behavioral evolution structure is more appropriate in the absence of a decision structure, which was the case developed in [145]; see Chapter 6. Here we shall ignore the behavioral part of the evolution structure, that is, we shall assume that $q^s_{s_1 \to s_2}$ may not vanish only for s_1, s_2 that differ only in their h-component. Moreover, since we will be doing analysis in the asymptotic regime of small interactions, it would be handy to introduce directly a small parameter δ_{int} discounting the power of interaction and thus denoting the corresponding rates by $\delta_{int} q^s_{s_1 \to s_2}$.

The above general structure can be very complicated. One can distinguish two natural simplifying assumptions: the set of edges is ordered and only the transitions between the neighbors are allowed, or the corresponding graph is complete, so that all transitions are allowed and have comparable rates. We shall choose the second alternative for B, and thinking about H as a hierarchy of agents, the first alternative for H. Moreover, we shall assume that the binary interaction occurs only within a common level in H (thus ignoring the binary interaction between the agents in different levels of the hierarchy). Hence for the transition rates $q_{i \to i+1,j}$ of the pressure structure increasing i we shall use the shorter notation q^+_{ij}, and for the transition rates $q_{i \to i-1,j}$ of the pressure structure decreasing i we shall use the shorter notation q^-_{ij}. Similarly, for the transition rates $\delta_{int} q^{ik}_{i \to i+1,j}$ of the hierarchical evolution structure increasing i we shall use the shorter notation $\delta_{int} q^{+k}_{ij}$, and for the transition rates $\delta_{int} q^{ik}_{i \to i-1,j}$ of the hierarchical evolution structure decreasing i we shall use the shorter notation $\delta_{int} q^{-k}_{ij}$.

Thus, if the elements of a matrix-valued function $x = (x_{ij})$ denote the occupations (densities or probabilities) of the states (i, j), and $u_{i,k \to j}$ (which may depend on time) is the decision matrix, the evolution of x is described by the ODE

$$\dot{x}_{ij} = \lambda \sum_{k \neq j} (u_{i,k \to j} x_{ik} - u_{i,j \to k} x_{ij}) + q^-_{i+1,j} x_{i+1,j} + q^+_{i-1,j} x_{i-1,j} - (q^+_{ij} + q^-_{ij}) x_{ij}$$

$$+ \delta_{int} \sum_{k \in B} (q^{+k}_{i-1,j} x_{i-1,k} x_{i-1,j} - q^{+k}_{ij} x_{ik} x_{ij}) + \delta_{int} \sum_{k \in B} (q^{-k}_{i+1,j} x_{i+1,k} x_{i+1,j} - q^{-k}_{ij} x_{ik} x_{ij}).$$

$$\tag{8.53}$$

It is assumed here that these equations hold for the internal points $i \neq 1, |H|$, and for the boundary values the terms containing the transition involving $i - 1$ or $i + 1$ respectively are omitted. In particular,

$$q^{+k}_{nj} = q^{-k}_{1j} = 0, \quad q^+_{nj} = q^-_{1j} = 0. \tag{8.54}$$

Explicit calculations can be essentially simplified under the additional constraint

$$q^+_{lj} = q^-_{l+1,j}, \tag{8.55}$$

for all l, j, which can be called the *detailed balance condition*: it asserts that the number of downgrades is compensated on average by the number of promotions.

To be able to identify the decision vector, we have to assign certain payoffs w_{ij} per unit time for remaining in the states (i, j) and the transition fees f^B_{kj} for transitions from (h, k) to (h, j) (which we assume independent of h for brevity) and the fines

f_j^H for transitions from (j, b) to $(j - 1, b)$ (which we assume independent on b for brevity). If $g_{ij} = g_{ij}(t)$ is the payoff to the state (i, j) in the process starting at time t and terminating at time T, and subject to a given evolution of x, these individual payoffs satisfy the following evolutionary HJB equation:

$$\dot{g}_{ij} = w_{ij} + \lambda \sup_u u_{i,j \to k}(g_{ik} - g_{ij} - f_{jk}^B) + q_{ij}^+(g_{i+1,j} - g_{ij}) + q_{ij}^-(g_{i-1,j} - g_{ij} - f_i^H)$$

$$+ \delta_{int} \sum_{k \in B} q_{ij}^{+k} x_{ik}(g_{i+1,j} - g_{ij}) + \delta_{int} \sum_{k \in B} q_{ij}^{-k} x_{ik}(g_{i-1,j} - g_{ij} - f_i^H). \quad (8.56)$$

As above, it is assumed here that these equations hold for the internal points $i \neq 1, |H|$, and for the boundary values the terms containing a transition involving $i - 1$ or $i + 1$ respectively are omitted. For instance, the equations with $i = 1$ are

$$\dot{g}_{1j} = w_{1j} + \lambda \sup_u u_{1,j \to k}(g_{ik} - g_{1j} - f_{jk}^B) + q_{1j}^+(g_{2j} - g_{1j}) + \delta_{int} \sum_{k \in B} q_{1j}^{+k} x_{ik}(g_{2j} - g_{1j}).$$

The *mean-field-game (MFG) consistency condition* means that one has to consider equations (8.53) and (8.56) on a given interval $[0, T]$ as a coupled system complemented by the initial condition x^0 for x and some terminal condition g_T for the optimal payoff g.

Assuming the discount δ_{dis} for future payoffs, one can look for a stationary optimal strategy, which leads to the stationary HJB equation

$$w_{ij} + \lambda \sup_u u_{i,j \to k}(g_{ik} - g_{ij} - f_{jk}^B) + q_{ij}^+(g_{i+1,j} - g_{ij}) + q_{ij}^-(g_{i-1,j} - g_{ij} - f_i^H)$$

$$+ \delta_{int} \sum_{k \in B} q_{ij}^{+k} x_{ik}(g_{i+1,j} - g_{ij}) + \delta_{int} \sum_{k \in B} q_{ij}^{-k} x_{ik}(g_{i-1,j} - g_{ij} - f_i^H) = \delta_{dis} g_{ij},$$
$$(8.57)$$

for every player in the state (i, j), and the fixed-point condition for evolution (8.53):

$$\lambda \sum_{k \neq j}(u_{i,k \to j} x_{ik} - u_{i,j \to k} x_{ij}) + q_{i+1,j}^- x_{i+1,j} + q_{i-1,j}^+ x_{i-1,j} - (q_{i,j}^+ + q_{i,j}^-)x_{ij}$$

$$+ \delta_{int} \sum_{k \in B}(q_{i-1,j}^{+k} x_{i-1,k} x_{i-1,j} - q_{ij}^{+k} x_{ik} x_{ij}) + \delta_{int} \sum_{k \in B}(q_{i+1,j}^{-k} x_{i+1,k} x_{i+1,j} - q_{ij}^{-k} x_{ik} x_{ij}) = 0.$$
$$(8.58)$$

The corresponding *stationary mean-field game (MFG) consistency condition* means that one has to consider equations (8.58) and (8.57) as a coupled stationary system.

In the limit of the fast execution of individual decisions $\lambda \to \infty$, the terms containing λ should vanish in (8.58) and (8.57) (if the execution is fast, in the stationary state no one should be interested in switching). Thus equations (8.57) and (8.58) become

$$w_{ij} + q_{ij}^+(g_{i+1,j} - g_{ij}) + q_{ij}^-(g_{i-1,j} - g_{ij} - f_i^H)$$

$$+ \delta_{int} \sum_{k \in B} q_{ij}^{+k} x_{ik}(g_{i+1,j} - g_{ij}) + \delta_{int} \sum_{k \in B} q_{ij}^{-k} x_{ik}(g_{i-1,j} - g_{ij} - f_i^H) = \delta_{dis} g_{ij},$$

$$\tag{8.59}$$

$$q_{i+1,j}^- x_{i+1,j} + q_{i-1,j}^+ x_{i-1,j} - (q_{i,j}^+ + q_{i,j}^-)x_{ij}$$

$$+ \delta_{int} \sum_{k \in B}(q_{i-1,j}^{+k} x_{i-1,k} x_{i-1,j} - q_{ij}^{+k} x_{ik} x_{ij}) + \delta_{int} \sum_{k \in B}(q_{i+1,j}^{-k} x_{i+1,k} x_{i+1,j} - q_{ij}^{-k} x_{ik} x_{ij}) = 0,$$

$$\tag{8.60}$$

supplemented by the consistency condition

$$g_{ik} - g_{ij} - f_{jk}^B \leq 0, \tag{8.61}$$

for all i, j, k such that $x_{ij} \neq 0$, ensuring that all u actually vanish in (8.59) for all occupied states.

As we mentioned, we look for the asymptotic regime with small evolution rates $\delta_{int} q_{ij}^{\pm k}$, that is, the terms involving these rates will be considered a perturbation to the linear system of two equations

$$w_{ij} + q_{ij}^+(g_{i+1,j} - g_{ij}) + q_{ij}^-(g_{i-1,j} - g_{ij} - f_i^H) = \delta_{dis} g_{ij} \iff (A_j^T + \delta_{dis})g._j = \tilde{w}._j,$$

$$\tag{8.62}$$

$$q_{i+1,j}^- x_{i+1,j} + q_{i-1,j}^+ x_{i-1,j} - (q_{i,j}^+ + q_{i,j}^-)x_{ij} = 0, \iff A_j x._j = 0, \tag{8.63}$$

supplemented by the consistency condition (8.61), where the matrix A_j is

$$A_j = \begin{pmatrix} q_{1j}^+ & -q_{2j}^- & 0 & \cdots & & \\ -q_{1j}^+ & (q_{2j}^+ + q_{2j}^-) & -q_{3j}^- & \cdots & & \\ & \cdots & & & & \\ & \cdots & & -q_{n-2,j}^+ & (q_{n-1,j}^+ + q_{n-1,j}^-) & -q_{nj}^- \\ & \cdots & & 0 & -q_{n-1,j}^+ & q_{nj}^- \end{pmatrix}, \tag{8.64}$$

A_j^T denotes its transpose, and $\tilde{w}_{ij} = w_{ij} - q_{ij}^- f_i^H$.

In this notation the full equation (8.59) takes the form

$$(A_j^T + \delta_{dis} - \delta_{int} E)g._j = \tilde{w}._j - \delta_{int} \sum_{k \in B} q_{ij}^{-k} x_{ik} f_i^H, \tag{8.65}$$

with the linear mapping E acting as

$$Eg_{.j} = \sum_{k \in B} q_{ij}^{+k} x_{ik}(g_{i+1,j} - g_{ij}) + \sum_{k \in B} q_{ij}^{-k} x_{ik}(g_{i-1,j} - g_{ij}). \qquad (8.66)$$

In the linear approximation (vanishing δ_{int}), the system of two equations become uncoupled. Moreover, different elements of B are also uncoupled, so that each of equations (8.62) and (8.63) can be solved separately for any j.

Let us solve system (8.63). Notice that this system is degenerate, as expected, since we are looking at positive solutions with $x_1 + \cdots + x_n = 1$. Thus the last equation can be discarded. Rewriting the system of the first $(n-1)$ equations using the first equation and then adding sequentially to each of the next $(n-2)$ equations the previous one yields the system

$$
\begin{aligned}
q_{1j}^{+} x_{1j} - q_{2j}^{-} x_{2j} &= 0, \\
q_{2j}^{+} x_{2j} - q_{3j}^{-} x_{3j} &= 0, \\
&\cdots \\
q_{n-1,j}^{+} x_{n-1,j} - q_{nj}^{-} x_{nj} &= 0.
\end{aligned}
\qquad (8.67)
$$

It has an obvious explicit solution, unique up to a multiplier:

$$x_{2j}^{*} = \frac{q_{1j}^{+}}{q_{2j}^{-}} x_{1j}^{*}, \quad x_{3j}^{*} = \frac{q_{2j}^{+}}{q_{3j}^{-}} \frac{q_{1j}^{+}}{q_{2j}^{-}} x_{1j}^{*}, \quad \cdots, \quad x_{nj}^{*} = \prod_{l=1}^{n-1} \frac{q_{lj}^{+}}{q_{l+1,j}^{-}} x_{1j}^{*}. \qquad (8.68)$$

Thus writing $x_j = \sum_i x_{ij}$, we obtain

$$x_{1j}^{*} = \left(1 + \frac{q_{1j}^{+}}{q_{2j}^{-}} + \frac{q_{2j}^{+}}{q_{3j}^{-}} \frac{q_{1j}^{+}}{q_{2j}^{-}} + \cdots + \prod_{l=1}^{n-1} \frac{q_{lj}^{+}}{q_{l+1,j}^{-}} \right)^{-1} x_{j}^{*}. \qquad (8.69)$$

Alternatively, starting the exclusion from the last equation of (8.67), we obtain

$$x_{n-1,j}^{*} = \frac{q_{nj}^{-}}{q_{n-1,j}^{+}} x_{nj}^{*}, \quad x_{n-2,j}^{*} = \frac{q_{n-1,j}^{-}}{q_{n-2,j}^{+}} \frac{q_{nj}^{-}}{q_{n-1,j}^{+}} x_{nj}^{*}, \quad \cdots, \quad x_{1j}^{*} = \prod_{l=1}^{n-1} \frac{q_{l+1,j}^{-}}{q_{l,j}^{+}} x_{nj}^{*},$$

$$(8.70)$$

$$x_{nj}^{*} = \left(1 + \frac{q_{nj}^{-}}{q_{n-1,j}^{+}} + \frac{q_{n-1,j}^{-}}{q_{n-2,j}^{+}} \frac{q_{nj}^{-}}{q_{n-1,j}^{+}} + \cdots + \prod_{l=1}^{n-1} \frac{q_{l+1,j}^{-}}{q_{l,j}^{+}} \right)^{-1} x_{j}^{*}. \qquad (8.71)$$

Thus we have proved that if either all $q_{ij}^{-} > 0$ or all $q_{ij}^{+} > 0$, the rank of matrix A_j is exactly $n - 1$, and the kernel of A is generated by the vector (8.68)–(8.69) or (8.70)–(8.71) respectively. Moreover, under the detailed balance condition (8.55), the matrix A is symmetric, and subject to the nondegeneracy of (i), its kernel is proportional to the uniform distribution: $x_{ij}^{*} = x_{j}^{*}/n$ for all i, j.

Though equation (8.62) can be solved explicitly, the exact formulas are extremely lengthy and can be better dealt with numerically. In order to grasp the main qualitative feature of the model, let us make a further simplifying assumption:

(A): The detailed balance conditions (8.55) hold with all their elements being strictly positive. In this case, we shall use the shorter notation

$$q_{ij} = q_{ij}^+ = q_{i+1,j}^-, \quad i = 1, \cdots, n - 1.$$

Then the zeroth-order approximation to the fixed point is proportional to the uniform distribution. Moreover, $A = A^T$, $Ker(A_j)$ is generated by the vector $(1, \cdots, 1)$, and the orthogonal complement to $Ker(A_j)$ is

$$Ker^\perp(A_j) = \{x : \sum x_j = 0\}.$$

Recall that we are looking for the asymptotic regime with small δ_{dis} (small discounting) and small δ_{int} (weak binary interaction). For a full analysis one has to make clear assumptions on the relation (*interaction-discounting relations*) between the small parameters δ_{dis} and δ_{int}. Three rough basic regimes can be naturally distinguished:

Assumption ID1: the interaction is very small: $\delta = \delta_{dis}$ and $\delta_{int} = \delta^2$;

Assumption ID2: the interaction and discounting are small effects of comparable order: $\delta = \delta_{dis} = \delta_{int}$;

Assumption ID3: the discounting is a stronger effect than the interaction: $\delta = \delta_{int}$ and $\delta_{dis} = \delta^2$.

A superficial analysis indicates several key common features of all these regimes. For instance, it turns out that in the leading order in small δ, the solution g^{0j} is uniform. This is an interesting effect showing that under all these regimes, the mixing is so strong that the combined effect of promotions and illegal behavior tend to equalize the incomes on all levels of the criminal hierarchy.

For example, let us concentrate on case ID1.

In the general position, the equation $A_j^T g_{.j} = \tilde{w}_{.j}$ has no solution, since the image of $A_j^T = A_j$ is $(n-1)$-dimensional, because the kernel $Ker(A_j)$ is one-dimensional. More precisely, the equation $A_j^T g_{.j} = \tilde{w}_{.j}$ has no solution if

$$(\tilde{w}_{.j}, x_{.j}^*) = \frac{1}{n} x_j^* \sum_i \tilde{w}_{ij} \neq 0. \tag{8.72}$$

Thus, in order to stay in the nondegenerate regime, the following assumption suggests itself:

(B): for all j,

$$\sum_i \tilde{w}_{ij} \neq 0. \tag{8.73}$$

Under (A) and (B), the solution to the equation (8.62) in the asymptotic limit of small δ should be of order $1/\delta$. Looking for the solution in the form of the expansion

$$g_{ij} = g^{0j}/\delta + g^{1j} + \delta g^{2j}$$

and equating the terms of order δ^{-1}, δ^0, δ^1, we obtain the equations

$$\begin{aligned} A_j g^{0j} &= 0, \\ A_j g^{1j} &= \tilde{w}_{\cdot j} - g^{0j}, \\ A_j g^{2j} - E g^{0j} + g^{1j} &= 0. \end{aligned} \tag{8.74}$$

The first equation tells us that g^{0j} belongs to the kernel $Ker\,(A_j)$, that is, $g^{0j}_\cdot = \alpha_j x^*_{\cdot j}$. As usual in the application of perturbation theory in the presence of degeneracy, to fully specify the leading-order asymptotics, one has to look at the equations of the next order. In fact, the second equation in (8.74) tells us that $\tilde{w}_{\cdot j} - g^{0j}$ belongs to the image of A, which coincides with $Ker^{\perp}(A_j)$:

$$\sum_i \tilde{w}_{ij} x^*_j / n = \alpha_j n (x^*_j/n)^2.$$

This implies that the main order of the small-δ asymptotics of the solution to (8.62) is g^{0j}/δ with

$$g^{0j}_i = \alpha_j x^*_{ij} = \alpha_j x^*_j / n = \sum_i \tilde{w}_{ij}/n. \tag{8.75}$$

Again the second equation in (8.74) does not specify g^{1j} uniquely. Looking at the third equation in (8.74) and noting that $E g^{0j} = 0$ for the uniform g^{0j}, we can conclude that g^{1j} belongs to the image of A_j, i.e., to $Ker^{\perp}(A_j)$. Thus to find g^{1j}, we have to invert A on $Ker^{\perp}(A_j)$. The direct calculations yield the following result: Under (A), let $y \in Ker^{\perp}(A_j)$. Then all solutions x to the equation $A_j x = y$ are given by the formula

$$\begin{cases} x_2 = x_1 - \dfrac{y_1}{q_{1j}}, \\[2ex] x_3 = x_1 - \dfrac{y_1}{q_{1j}} - \dfrac{y_1 + y_2}{q_{2j}}, \\[1ex] \cdots \\[1ex] x_n = x_1 - \dfrac{y_1}{q_{1j}} - \cdots - \dfrac{y_1 + \cdots + y_{n-1}}{q_{n-1,j}}, \end{cases} \tag{8.76}$$

with arbitrary x_1. There exists a unique solution $x \in Ker^{\perp}(A_j)$ specified by

$$x_1 = \frac{n-1}{n}\frac{y_1}{q_{1j}} + \frac{n-2}{n}\frac{y_1+y_2}{q_{2j}} + \cdots + \frac{1}{n}\frac{y_1+\cdots+y_{n-1}}{q_{n-1,j}}. \tag{8.77}$$

Formulas (8.76), (8.77) yield $x = g^{1j}$ when

$$y_i = \tilde{w}_{ij} - g_i^{0j} = \tilde{w}_{ij} - (\delta_{int}/\delta)\sum_i \tilde{w}_{ij}/n + (\delta_{int}/\delta)\sum_{k\in B} q_{ij}^{-k} x_{ik} f_i^H.$$

Again in the main order in δ, the consistency condition (8.61) can be written as

$$\sum_i \tilde{w}_{ik} - \sum_i \tilde{w}_{ij} = O(\delta). \tag{8.78}$$

Not assuming that w depends on δ, this leads to the remarkable result that in the equilibrium of the asymptotic regime of small δ, only those levels $j \in B$ are occupied (that is, $x_j^* \neq 0$) where the sum $\sum_i \tilde{w}_{ij}$ takes its maximum.

Thus, in order to stay in general position, another assumption suggests itself: (C): there exists $b \in B$ such that

$$\sum_i \tilde{w}_{ib} > \sum_i \tilde{w}_{ij}, \quad j \neq b. \tag{8.79}$$

In this case, the solution to (8.62), (8.63), (8.61) to first order in small δ is given by

$$x_{ib}^0 = 1/n, \; x_{ij}^0 = 0 \text{ for all } j \neq b \text{ and all } i; \quad g_{ib} = \delta^{-1}\sum_i \tilde{w}_{ib}/n. \tag{8.80}$$

More strongly, assumption (C) implies that in every equilibrium x^* with δ sufficiently small,

$$x_{ij}^* = 0 \text{ for all } j \neq b \text{ and all } i, \tag{8.81}$$

so that all terms with $j \neq b$ become irrelevant for the analysis.

Extending (C), one has the following assumption:

$(\tilde{C})$: there are several points $b_1, \cdots, b_l$ of maxima of $\sum_i \tilde{w}_{ib}$.

In this case, one cannot distinguish the states with these b_k to first order in δ, and thus we have a $(l-1)$-parameter family of solutions parametrized by nonnegative numbers $x_{b_l}^*$ normalized to the condition

$$\sum_{k=1}^l x_{b_k}^* = 1.$$

Thus under (C), the equilibrium solution remains in b, and in case of $(\tilde{C})$, it can oscillate among $b_1, \cdots, b_l$. In any case, one can explicitly calculate these stationary solutions and show that they form turnpikes, extending results of Section 8.3 (see details in [126]).

Speaking more generally, the model of this section can be looked at as an extension of the model of Section 3.5. Namely, instead of a family of rates $Q_{ij}(x, b)$, we have now a family $Q_{ij}^k(x, b)$ with an additional index k that can be chosen strategically by the small players. Under a nondegeneracy assumption of type (C) above, the small players will choose k that would yield the most effective turnpike among those described by Theorems 3.5.1 and 3.5.2.

Appendix
Notes and Compliments

As a general reference for the analysis of multiagent systems we refer to [198, 215]. The development of multiagent games and multiagent systems can be considered a part of big data science, contributing to the understanding of smart technologies, the systems of systems, and the internet of things.

The literature on both our main methodology (dynamic law of large numbers) and the concrete areas of applications (inspection, corruption, etc.) is quite abundant and continues to grow rapidly. In this section we provide an overview of at least the main trends most closely related to our methods and objectives.

1. *Dynamic LLN for Optimal Control.*

The dynamic law of large numbers (LLN) for interacting particles is a well-developed field of research in statistical mechanics and stochastic analysis; see, e.g., [136] for a review, and for the kinetic models, see, e.g., [8] and references therein. For agents, that is, in the framework of controlled processes and games, the LLN was strongly advanced in the setting of evolutionary biology, leading to a deep analysis of the replicator dynamics of evolutionary games and its various versions and modifications (see, e.g., [115] or [144] for a general picture). A random multiagent approximation for a given dynamics can be constructed in various ways. For instance, for the classical replicator dynamics, a multiagent approximation was analyzed via migrations between the states preserving the total number of agents (see, e.g., [37]) or via the model arising from the standard biological interpretation of payoffs as the reproduction rate (thus changing the number of agents; see the last chapter in [144]).

For the model in which pairwise interaction is organized in discrete time so that at any moment, a given fraction $\alpha(N)$ of a homogeneous population of N species is randomly chosen and decomposed into matching pairs, which afterward experience simultaneous transformations into other pairs according to a given distribution, the convergence to a deterministic ODE is proved in [50]. The paper [68] extends this setting to include several types of species and the possibility of different scalings that may lead, in the limit $N \to \infty$, not only to an ODE, but to a diffusion

© Springer Nature Switzerland AG 2019

V. N. Kolokoltsov and O. A. Malafeyev, *Many Agent Games in Socio-economic Systems: Corruption, Inspection, Coalition Building, Network Growth, Security*, Springer Series in Operations Research and Financial Engineering, https://doi.org/10.1007/978-3-030-12371-0

process. In [130], the general class of the stochastic dynamic law of large numbers is obtained from binary or more general kth-order interacting particle systems (including jump-type and Lévy processes as a noise). The study [37] concentrates on various subtle estimates for the deviation of the limiting deterministic evolution from the approximating Markov chain for the evolution that allows a single player (at any random time) to change its strategy to the strategy of another randomly chosen player. Namely, it is shown, using the methods developed in [88], that the probability that the maximum deviation of the trajectories of the Markov chain with N players from the limiting evolution exceeds some ϵ is bounded by the exponent $2me^{-\epsilon^2 cN}$, where m is the number of states and c some constant.

A related trend of research analyzes various choices of Markov approximation to repeated games and their consequences to the question of choosing a particular Nash equilibrium among the usual multitude of them. The seminal contribution [124] distinguishes specifically the myopic hypothesis, the mutation or experimentation hypothesis, and the inertia hypothesis in building a Markov dynamics of interaction. As shown in [124] (with a similar result in [234]), introducing the mutation of strength λ and then passing to the limit $\lambda \to 0$ allows one to choose a certain particular Nash equilibrium, called a long run equilibrium (or statistically stable, in the terminology of [87]), that for some coordination games turns out to coincide with the risk-dominant (in the sense of [111]) equilibrium. Further important contributions in this direction include [46, 47, 79], showing how different equilibria could be obtained by a proper fiddling with noise (for instance local or uniform as in [79]) and discussing the important practical question "how long is the long run?" (for recent progress on this question see [156]). In particular, the paper [47] exploits the model of "musical chairs," whereby changes in strategies occur according to a model that is similar to the children's game in which players outnumber chairs, and those unable to find a chair when the music stops must leave the game. This paper discusses in detail the crucial question of the effect of applying the limits $t \to \infty, \tau \to 0$ (the limit from discrete to continuous replicator dynamics), $N \to \infty$ and $\lambda \to 0$ in various orders. The further development of the idea of local interaction leads naturally to the analysis of the corresponding Markov processes on large networks; see [166] and references therein. Some recent general results of the link between Markov approximation to the mean-field (or fluid) limit can be found in [39, 161]. Though in many papers on Markov approximation, the switching probabilities of a revising player depend on the current distribution of strategies used (assuming implicitly that this distribution is observed by all players), there exist also interesting results (initiated in [209]; see new developments in [210]) arising from the assumption that the switching of a revising player is based on an observed sample of given size of randomly chosen other players. Let us also mention the papers [91, 91], devoted to refined mean-field approximations, where the next terms are derived by assuming higher regularity of the coefficients.

A slightly different (but still very close) trend of research is represented by the analysis of general stochastic approximation in association with the so-called method of ordinary differential equations; see, e.g., [38, 204] and references therein.

The convergence results for a centrally controlled Markov chain of a large number of constituents to the deterministic continuous-time dynamics given by ordinary differential equations (kinetic equations), dealt with in detail in our Chapters 2, 3, were apparently initiated in papers [90, 138] from two different approaches on the level of trajectories (in the spirit of the paper [37]) and for the averages via the semigroup methods, as we do it here. In the paper [63], the theory of coalition-building was developed via the method of [90]. An extension of the method of [138] is given in [20].

The analysis of kinetic equations with fractional derivatives was initiated in [127]. However, the setting (and the equations discussed) of this paper is different from that of our Section 2.14.

Proving the convergence of the process of fluctuations of multiagent or multiparticle systems from their dynamic LLN, that is, the dynamic central limit theorem (CLT), is of course harder than getting the LLN itself. In fact, proving such a CLT for the model of coagulation (or merge and splitting) of Chapter 4 was specifically mentioned as an important open problem in the influential review [5], the solution being provided in [134] based on the general analytic methodology [131]. The latter methodology can also be directly applied to the finite-state models of this book.

2. Mean-field Games (MFGs).

Mean-field games, which are dealt with in the second part of this book, present a quickly developing area of the game theory. Mean-field games were initiated by Lasry–Lions [159] and Huang–Malhame–Caines [118, 119]. Roughly speaking, they deal with situations involving many players with similar interests when the influence of each player on the overall outcome becomes negligible as the number of players tends to infinity. As insightful illustrative examples for the theory, one can mention two problems (suggested by P.-L. Lions): (1) "when does the meeting start?" whereby agents choose the time to plan to arrive at a meeting that should begin when a certain fixed number of participants arrive, and where costs for participants arise from being too early and too late (a game of time); and (2) "where did I put my towel on the beach?" in which agents try to choose a place on the beach as close as possible to some point of interest (restaurant or seashore) but not too close to other agents (a congestion game).

At present, there exist several excellent surveys and monographs on various directions of the theory; see [28, 40, 54, 60, 99, 109]. For other key recent developments, let us mention the papers [21, 29, 57, 58, 61, 149, 220], and references therein. Let us mention specifically the development of mean-field games with a major player; see [120, 159, 186, 228], where also the necessity to consider various classes of players is well recognized; see also [42, 59].

The papers [95, 96], as well as [105, 105], initiated the study of finite-state-space mean-field games that are the objects of our analysis in the second part of this book. The paper [95] develops the theory of discrete-time MFGs, showing, in particular, the convergence of the solutions of the backward–forward MFG consistency problem to a stationary problem, thus initiating a very fruitful discussion of the precise links between stationary and time-dependent solutions. The paper [96] deals with more conventional continuous-time models, proving results on the existence and uniqueness of the solutions and the convergence of Nash equilibria of approximating

N-player games to a solution of the MFG consistency problem. The papers [105, 108] are devoted specifically to state spaces that are graphs with a nontrivial geometry. All these finite-state models (their formulations and results) are based on a rather restrictive structure of control and payoff (for instance, the transition rates are linear functions of a control parameter, strong convexity assumptions for payoffs are assumed), which also yield a quite specific form of the master equation, which in some cases (for the so-called potential MFGs) is even reduced to a hyperbolic system of quasilinear PDEs. For models with more general dependence of dynamics and payoffs on the control parameters, the analysis of MFGs was initiated in the papers [30, 145, 146], developed further in this book. Of course, the simplest MFGs arise from the two-state models, for which many things can be calculated explicitly; see, e.g., the papers [43, 98], where these two-state games are applied to the analysis of socioeconomic models of paradigm shifts in scientific research and of consumer choices of products whose price depends on consumption rates. A setting for stationary problems with a finite lifetime of agents (and thus with a varying number of agents) was suggested and analyzed in [230].

The applications of mean-field games are developing rapidly. For instance, [3, 73] analyze the problem of evacuation in a crowded building or room (say, in case of a fire); opinion dynamics in social networks is analyzed in [32]. The paper [22] deals with the demand management of electrical heating or cooling appliances providing desynchronization of agents using a bang-bang switching control. The paper [23] deals with the dynamic of characteristic functions (values of coalitions) in transferable utility coalition games. Optimal stopping games are dealt with in [189]. Examples with more economics content include the standard models of the exploitation of common resources and the formation of prices that influence total sales (see, e.g., [60, 107] and references therein). As an example with a biological flavor, one can mention the problem of flocking of birds; see [60, 186].

Some preliminary ideas pointing in the direction of the MFG structure can be traced back to the paper [123] and references therein.

3. Corruption and Inspection Games.

The analysis of the spread of corruption in bureaucracies is a recognized area of the application of game theory that has attracted the attention of many researchers. General surveys can be found in [4, 122, 163]; see also the monographs [200, 201]. In his prize lecture [121], L. Hurwicz gives an introduction in layman's terms to various problems arising in an attempt to find out "who will guard the guardians?" and what mechanisms can be exploited to enforce legal behavior?

In a series of papers [157, 158], the authors analyze the dynamic game in which entrepreneurs have to apply to a set of bureaucrats (in a prescribed order) to obtain approval for their business projects. Before granting approval, the bureaucrats ask for bribes, with the amounts of the bribes considered strategies for the bureaucrats. This model is often referred to as petty corruption, since each bureaucrat is assumed to ask for a small bribe, so that the large bureaucratic losses of entrepreneurs occur from a large number of bureaucrats. This is an extension of the classical ultimatum game, because the game stops whenever an entrepreneur declines to pay the required graft. The existence of an intermediary undertaking the contacts with bureaucrats for a fee may essentially affect the outcomes of this game.

In the series of works [182, 225, 226], the authors develop a hierarchical model of corruption, whereby the inspectors of each level audit the inspectors of the previous level and report their findings to the inspector of the next upper level. For a the payment of a bribe they may choose to make a falsified report. The inspector at the highest level is assumed to be honest but very costly for the government. The strategy of the government is in the optimal determination of the audits on various levels with the objective of achieve the minimal level of corruption with minimal cost. Related hierarchical models are developed in [102, 103], where the stress is on the interaction of the three types of players: benevolent dictator (the government), corrupt bureaucrat, and an agent (a producer of goods), and in particular, on the conditions allowing for the stable development of the economy

The paper [217] develops a simple model to obtain insight into the problem of when unifying efforts result in strengthening of corruption. In the paper [171], the model of a network corruption game is introduced and analyzed, with the dynamics of corrupt services between entrepreneurs and corrupt bureaucrats propagating via a chain of intermediaries. In [181], the dichotomy between public monitoring and governmental corruptive pressure on the growth of an economy was modeled. In [162], an evolutionary model of corruption is developed for ecosystem management and biodiversity conservation.

In [128], a model of corruption with regard to psychological mimicry in an administrative apparatus with three forms of corruption is constructed. It is given in terms of a system of four differential equations describing a number of different groups. The equilibrium states that make it possible to specify the dominant form of corruption and investigate its stability, depending on the parameters of the psychological mimicry and the rigor of anticorruption laws, are found. In [172], the corruption dynamics is analyzed by means of a lattice model similar to the three-dimensional Ising model: agents placed at the nodes of a corrupt network periodically choose to perform or not to perform an act of corruption. In [173], the transportation problem of multiagent interaction between transporters of different goods with a corrupt component is introduced and studied. A statistical procedure of anticorruption control of economic activity is proposed in [195]. A model of optimal allocation of resources for anticorruption purposes is developed in [180]. Various approaches to modeling corruption are collected in the monograph [170].

The research on the political aspects of corruption develops around Acton's dictum that "power corrupts," where elections usually serve as a major tool of public control; see [94] and references therein.

Closely related are the so-called inspection games; see the surveys, e.g., in [18, 125, 144]. Our evolutionary approach was initiated in [148], and similar ideas can be found in [106]. Inspection games model noncooperative interactions between two strategic parties, called the inspector and inspectee. The inspector aims to verify that certain regulations imposed by the benevolent principal for whom the inspector is acting are not violated by the inspectee. In contrast, the inspectee has an incentive to disobey the established regulations, risking the enforcement of a punitive fine in the case of detection. The introduced punishment mechanism is a key element of inspection games, since deterrence is generally the inspector's highest priority. Typically, the inspector has limited means of inspection at his/her disposal, so that his/her detection efficiency can only be partial.

One of the first models was a two-person zero-sum recursive inspection game proposed by Dresher [76], where it was supposed that a given number n of periods are available for an inspectee to commit, or not, a unique violation, and a given number $m \leq n$ of one-period lasting inspections are available for the inspector to investigate the inspectees' adherence to the rules, assuming that a violator can be detected only if caught (inspected) in the act. Important extensions were given in the papers [71, 117]. This work initiated the application of inspection games to arms control and disarmament; see [16] and references therein. This basic model was generalized in [174] to a non-zero-sum game, adopting the notion of inspector leadership and showing (among other things) that the inspector's option to announce and commit to a mixed inspection strategy in advance actually increases his/her payoff.

In [221], a similar framework was applied to investigate the problem of a patroller aiming to inhibit a smuggler's illegal activity. In their so-called customs–smuggler game, customs patrol uses a speedboat to detect a smuggler's motorboat attempting to ship contraband through a strait. They introduced the possibility of more than one patrolling boat, namely the possibility of two or more inspectors, potentially not identical, and suggested the use of linear programming methods for the solution of such scenarios. Closed-form solutions for the case of two patrolling boats and three patrolling boats were provided in [31, 89] respectively. This research initiated the flood of literature on so-called *patrolling games*; see [9, 10] for further developments.

In a series of papers, Von Stengel (see [227] and references therein) introduced a third parameter in Dresher's game, allowing multiple violations, but proving that the inspector's optimal strategy is independent of the maximum number of the inspectees' intended violations. He studied several variations, (i) optimizing the detection time of a unique violation that is detected at the following inspection, given that an inspection is not currently taking place, (ii) adding different rewards for the inspectees' successfully committed violations, and (iii) extending Maschler's inspector leadership version under the multiple intended violations assumption. The papers [82, 207] studied a similar three-parameter, perfect-capture sequential game, where (i) the inspectee has the option to legally violate at an additional cost; (ii) a detected violation does not terminate the game; (iii) every uninspected violation is disclosed to the inspector at the following stage.

Non-zero-sum inspection became actively studied in the 1980s, in the context of the nuclear nonproliferation treaty (NPT). The perfect-capture assumption was partly abandoned, and errors of Type 1 (false alarm) and Type 2 (undetected violation given that inspection takes place) were introduced to formulate the so-called imperfect inspection games; see, e.g., [55] and references therein for the solution of imperfect, sequential, and nonsequential games, assuming that players ignore any information they collect during their interactions, whereby an illegal action must be detected within a critical timespan before its effect is irreversible. Imperfect inspection and timely detection in the context of environmental control were developed in [205].

Avenhaus and Kilgour [17] developed a non-zero-sum, imperfect (Type 2 error) inspection game, wherein a single inspector can continuously distribute his/her effort/resources between two noninteracting inspectees, exempted from the simplistic dilemma of whether to inspect or not. They related the inspector's detection efficiency with respect to the inspection effort through a nonlinear detection function and derived results for the inspector's optimum strategy subject to its convexity. The paper [116] extended this model, considering a similar $(N + 1)$-player inspection game, where the single inspection authority not only intends to optimally distribute his/her effort among N inspectee countries, but also among several facilities within each inspectee's country. These and related models are presented in detail in the foundational monograph [19].

The methodology of the dynamic LLN, which we develop in this book, can be applied in fact to almost all these models, allowing one to deal effectively with situations in which many agents take part and some bulk characteristics of the dynamics of the game are of interest.

4. Security and Bioterrorism.

Game-theoretic papers dealing with counterterrorism modeling were briefly mentioned in Section 1.8. Our approaches with the LLN makes the bridge between these papers and another trend of research, in which the development of the extremists' activity is modeled by some system of ODEs of low dimension, which present variations of the models of the propagation of infectious diseases in epidemiology. The standard abbreviations for the population classes in the latter theory are S for susceptible, L or E for latent or exposed (infected but yet not infectious), I for infectious, and R for recovered individuals. Depending on which classes are taken into account, several standard models were developed, including SI epidemics, SEI epidemics, SIR epidemics, and $SEIR$ (or $SLIR$ in other notations) epidemics. For instance, the latter model studies the process of the evolution of the number of classes S, E, I, R under the intuitively very meaningful system of four ODEs

$$\dot{S} = -\lambda SI, \quad \dot{L} = \lambda XI - \alpha L, \quad \dot{I} = \alpha L - \mu I, \quad \dot{R} = \mu I$$

with some constants α, μ, λ, and the SIR model is obtained by replacing the middle two equations by the equation $\dot{I} = \lambda XI - \mu I$; see the reviews, e.g., in [114] or [197] for various modifications.

Modifying this class of models to the propagation of extremism, the authors of [62] group the population into four classes: fanatics F, semifanatics (not fully converted) E, susceptible S, and raw material (indifferent) G. With some reasonable assumption on the propagation of influences via binary interaction, the authors build a system of ODEs in four variables with a quadratic r.h.s. (depending on certain numeric coefficients) and study its stationary points. In [213], devoted concretely to modeling ETA (Basque nationalist organization) in Spain, the authors distinguish the following classes of population: those against independence, E, those striving for independence but without violence, N, those supporting the fight for independence with violence, V, and the rest, A. Again a reasonable system of ODEs in four variables is built, and its predictions are compared with real figures in an attempt to evaluate the key parameters. In [104], the authors group the population into three classes: extremists E, susceptible S, and reserved (isolated in jails). Moreover, they add an additional variable G measuring the level of effort of the government G, and build and study a system of ODEs in four variables (depending on certain numeric coefficients). The paper [223] builds a model with three variables, the number of terrorists T, the number of susceptible (to terrorist ideas) S, and the number of unsusceptible N, where the efforts of society are distributed between two types of interference: direct military or police intervention (the analogue of the preemptive measures of [202] discussed in Section 1.8) decreasing T, and propaganda or concessions decreasing S, with the usual transitions between the groups via binary interactions. The system of ODEs suggested in [223] looks like

$$\dot{T} = aTS - bT^2 + cT, \quad \dot{S} = -aTS - eT^2S + fT + gS, \quad \dot{N} = eT^2S - f_1T + hN,$$

with constants a, b, c, e, f, g, f_1, h. The less-intuitive quadratic and third-order terms with b and e reflect two methods above of the influence of society. The stationary points of this dynamics are classified depending on the "control parameters" b and e. The paper [214] builds a model with terrorist groups of size T_1 and T_2 acting in two neighboring countries with counterterrorist measures in those countries, N_1, N_2, measured in some units, the corresponding system of ODEs being

$$\dot{T_1} = -a_1N_1T_1 - b_1\dot{T_1} + g_{12}T_2, \quad \dot{T_2} = -a_2N_2T_2 - b_2\dot{T_2} + g_{21}T_1, \quad \dot{N_1} = -\gamma_1 T_1, \quad \dot{N_2} = -\gamma_2 T_2.$$

We have shown just a few examples of models. Their variety reveals that mathematical modeling of these very complex processes is effectively in its initial phases, without any consensus about the relevant parameters and how they should be estimated.

All these models can be enhanced by our methodology by linking their evolutions with more real multiagent models and by considering the efforts of the government as a strategic parameter (in the spirit of Chapter 3) or including some optimization objective for the individuals of each group (in the spirit of Chapter 1 or 6), leading to the estimates of the investments needed to control the situation.

An extensive introduction to various approaches for analyzing terrorism can be found in [80], including statistical parameter estimates, linear regression, utility minimization on $\mathbf{R}_+^2$ (for the choice between proactive and defensive measures), and several game-theoretic formulations, for example as social dilemmas (static two-player games) or as dynamic modeling in the spirit of the entry deterrence games. The latter consider the question of committing or not a terrorist attack under the same setting as entering or not the market occupied by a monopolist.

Let us also mention the paper [231] (and several references therein), which uses the methods of deterministic optimal control to analyze governmental investments in the detection and interdiction of terrorist attacks. The main model of this paper looks at the evolution of the number of undetected X and detected Y plots of attacks subject to the ODEs

$$\dot{X} = \alpha - \mu X - \delta(f - Y)X, \quad \dot{Y} = \delta(f - Y)X - \rho Y,$$

where α and f are the control parameters of terrorists and the government respectively.

An extensive empirical analysis of crime models from the point of view of statistical physics is developed in [190, 191].

Similar models can be used for the analysis of the propagation of scientific ideas. The paper [36] uses all the basic versions of deterministic epidemics presented above (SEI, SIR, $SEIR$) to assess the propagation of the method of Feynman diagrams through the community of physicists, a method that had many influential opponents at its initiation, such as Oppenheimer in the USA and Lev Landau in the USSR. Yet another twist of the problem concerns the illegal consumption of energy resources; see [6] and references therein.

Recently, more attention has been paid to models of cybersecurity, as was discussed in Section 1.8 with reference to [41] for a review.

5. Optimal resource allocation.

Approaches to the optimal allocation of resources are numerous and are applied in various frameworks. For instance, one can distinguish resource allocation in the framework of reliability theory, where one looks for the optimal amount of redundant units to sustain the work of complex systems without interruption (see, e.g., [224]), the approaches based on the mechanism design methodology (distinguishing market and nonmarket mechanisms for allocations; see, e.g., [67] and references therein) and on Bayesian statistical inference (see, e.g., [236] and references therein). An approach from the theory of swarm intelligence (see [48]) models the allocation of tasks and resources by analogy with the procedures found in insect communities (ants or bees) and market-based control by analogy with market bidding–clearing mechanisms (see [66, 168]). Closely related are questions of the exploitation of limited common resources, often formulated as fish wars, (see, e.g., [154, 175, 176]), project management problems (see, e.g., [183]), distribution of investments between economic sectors (see [219]), and general control of crowd behavior and information propagation through crowds (see [26, 53] and references therein). Our approach to optimal allocation, taken from [140], deals with the distribution of the efforts of the principal for better management of mean-field interacting particle systems and is also close in spirit to [188].

References

[1] J.A. Acebrón, L.L. Bonilla, J. Perez Vicente, F. Riort, R. Spiegler, The kuramoto model: a simple paradigm for synchronization phenomena. Rev. Mod. Phys. **77**, 137–185 (2005)

[2] Y. Achdou, F. Camilli, I. Capuzzo-Dolcetta, Mean field games: convergence of a finite difference method. SIAM J. Numer. Anal. **51**(5), 2585–2612 (2013)

[3] Y. Achdou, M. Laurier, On the system of partial differential equations arising in mean field type control. Discret. Contin. Dyn. Syst., A **35**, 3879–3900 (2015)

[4] T.S. Aidt, Economic analysis of corruption: a survey. Econ. J. **113**(491), F632–F652 (2009)

[5] D.J. Aldous, Deterministic and stochastic models for coalescence (aggregation and coagulation): a review of the mean-field theory for probabilists. Bernoulli **5**(1), 3–48 (1999)

[6] F.T. Aleskerov, B.M. Schit, Production, legal and illegal consumption of electric energy (in Russian). Probl. Control (Problemy upravlenia) **2**, 63–69 (2005)

[7] G.V. Alferov, O.A. Malafeyev, A.S. Maltseva, Programming the robot in tasks of inspection and interception, in *Proceedings of IEEE 2015 International Conference on mechanics. Seventh Polyakov's Reading* (Petersburg, Russia, 2015), pp. 1–3

[8] J. Almquist, M. Cvijovic, V. Hatzimanikatis, J. Nielsen, M. Jirstrand, Kinetic models in industrial biotechnology improving cell factory performance. Metab. Eng. **24**, 38–60 (2014)

[9] S. Alpern, Th Lidbetter, Mining coal or finding terrorists: the expanding search paradigm. Oper. Res. **61**(2), 265–279 (2013)

[10] S. Alpern, A. Morton, K. Papadaki, Patrolling games. Oper. Res. **59**(5), 1246–1257 (2011)

[11] W.J. Anderson, Continuous -time markov chains, in *Probability and its Applications*, Springer Series in Statistics (Springer, Berlin, 1991)

[12] L. Andreozzi, Inspection games with long-run inspectors. Eur. J. Appl. Math. **21**(4–5), 441–458 (2010)

[13] D. Arce, T. Sandler, Counterterrorism: a game-theoretic analysis. J. Confl. Resolut. **49**, 183–200 (2005)

[14] R. Avenhaus, Applications of inspection games. Math. Model. Anal. **9**, 179–192 (2004)

[15] R. Avenhaus, M.J. Canty, Playing for time: a sequential inspection game. Eur. J. Oper. Res. **167**(2), 475–492 (2005)

[16] R. Avenhaus, M.D. Canty, D.M. Kilgour, B. von Stengel, S. Zamir, Inspection games in arms control. Eur. J. Oper. Res. **90**(3), 383–394 (1996)

[17] R. Avenhaus, D. Kilgour, Efficient distributions of arm-control inspection effort. Nav. Res. Logist. **51**(1), 1–27 (2004)

© Springer Nature Switzerland AG 2019

V. N. Kolokoltsov and O. A. Malafeyev, *Many Agent Games in Socio-economic Systems: Corruption, Inspection, Coalition Building, Network Growth, Security*, Springer Series in Operations Research and Financial Engineering, https://doi.org/10.1007/978-3-030-12371-0

[18] R. Avenhaus, B. Von Stengel, S. Zamir, Inspection games, in *Handbook of Game Theory with Economic Applications*, vol. 3, ed. by R. Aumann, S. Hart (North-Holland, Amsterdam, 2002), pp. 1947–1987

[19] R. Avenhaus, T. Krieger, Inspection games over time - Fundamental models and approaches, Monograph to be published in 2019

[20] Yu. Averboukh, Extremal shift rule for continuous-time zero-sum Markov games. Dyn. Games Appl. **7**(1), 1–20 (2017)

[21] Yu. Averboukh, A minimax approach to mean field games. Mat. Sbornik **206**(7), 332 (2015)

[22] F. Bagagiolo, D. Bauso, Mean-field games and dynamic demand management in power grids. (English summary). Dyn. Games Appl. **4**(2), 155–176 (2014)

[23] D. Bauso, T. Basar, Strategic thinking under social influence: scalability, stability and robustness of allocations. Eur. J. Control **32**, 115 (2016)

[24] J.M. Ball, J. Carr, The discrete coagulation-fragmentation equations: existence, uniqueness, and density conservation. J. Stat. Phys. **61**(1/2), 203–234 (1990)

[25] J.M. Ball, J. Carr, O. Penrose, The Becker-Dring cluster equations: basic properties and asymptotic behaviour of solutions. Commun. Math. Phys. **104**(4), 657–692 (1986)

[26] N. Barabanov, N.A. Korgin, D.A. Novikov, A.G. Chkhartishvili, Dynamic models of informational controlin social networks. Autom. Remote. Control. **71**(11), 2417–2426 (2010)

[27] A.-L. Barabási, R. Albert, Emergence of scaling in random networks. Science **286**, 509–512 (1999)

[28] M. Bardi, P. Caines, I. Capuzzo Dolcetta, Preface: DGAA special issue on mean field games. Dyn. Games Appl. **3**(4), 443–445 (2013)

[29] R. Basna, A. Hilbert, V. Kolokoltsov, An epsilon-Nash equilibrium for non-linear Markov games of mean-field-type on finite spaces. Commun. Stoch. Anal. **8**(4), 449–468 (2014)

[30] R. Basna, A. Hilbert, V.N. Kolokoltsov, An approximate Nash equilibrium for pure jump Markov games of mean-field-type on continuous state space Stochastics. **89**(6–7), 967–993 (2016). http://arxiv.org/abs/1605.05073

[31] V.J. Baston, F.A. Bostock, Naval. Res. Logist. **38**, 171–182 (1991)

[32] D. Bauso, H. Tembine, T. Basar, Opinion dynamics in social networks through mean-field games. SIAM J. Control Optim. **54**(6), 32253257 (2016)

[33] H. Beecher Stowe, *Uncle Tom's Cabin* (Blackie, 1963)

[34] G.S. Becker, G.J. Stigler, Law enforcement, malfeasance, and compensation of enforces. J. Leg. Stud. **3**(1), 1–18 (1974)

[35] V.P. Belavkin, V. Kolokoltsov, On general kinetic equation for many particle systems with interaction, fragmentation and coagulation. Proc. R. Soc. Lond. A **459**, 727–748 (2003)

[36] L.M.A. Bettencourt, A. Cintron-Arias, D.I. Kaiser, C. Castillo-Chavez, The power of a good idea: Quantitative modeling of the spread of ideas from epidemiological models. Phys. A **364**, 513536 (2006)

[37] M. Benaim, J. Weibull, Deterministic approximation of stochastic evolution in games. Econometrica **71**(3), 873–903 (2003)

[38] M. Benaim, A dynamical system approach to stochastic approximations. SIAM J. Control Optim. **34**(2), 437–472 (1996)

[39] M. Benaim, J.-Y. Le Boudec, A class of mean field interaction models for computer and communication systems. Perform. Eval. **65**, 823–838 (2008)

[40] A. Bensoussan, J. Frehse, P. Yam, *Mean Field Games and Mean Field Type Control Theory* (Springer, Berlin, 2013)

[41] A. Bensoussan, S. Hoe, M. Kantarcioglu, A game-theoretical approach for finding optimal strategies in a botnet defense model, in *Decision and Game Theory for Security First International Conference, GameSec 2010*, vol. 6442, ed. by T. Alpcan, L. Buttyan, J. Baras (Springer, Berlin, 2010), pp. 135–148

[42] A. Bensoussan, J. Frehse, The Master equation in mean field theory. Journal de Mathématiques Pures et Appliquées, (9) **103**(6), 1441–1474 (2015)

[43] D. Beancenot, H. Dogguy, Paradigm shift: a mean-field game approach. Bull. Econ. Res. **67**(3), 0307–3378 (2015)

[44] M. Bell, S. Perera, M. Piraveenan, M. Bliemer, T. Latty, Ch. Reid, Network growth models: A behavioural basis for attachment proportional to fitness. Sci. Rep. **7** (2017). Article number 42431

[45] T. Besley, J. McLaren, Taxes and bribery: the role of wage incentives. Econ. J. **103**(416), 119–141 (1993)

[46] K. Binmore, L. Samuelson, Muddling through: noisy equilibrium selection. J. Econ. Theory **74**(2), 235–265 (1997)

[47] K. Binmore, L. Samuelson, R. Vaughan, Musical chairs: modeling noisy evolution. Games Econ. Behav. **11**(1), 1–35 (1995)

[48] E. Bonabeau, M. Dorigo, G. Theraulaz, *Swarm Intelligence. From Natural to Artificial Systems*, Santa Fe Institute Studies in the Sciences of Complexity (Oxford University Press, Oxford, 1999)

[49] S. Bowles, *Microeconomics. Behavior, Institutions and Evolution* (Russell Sage Foundation, 2004)

[50] R. Boylan, Continuous approximation of dynamical systems with randomly matched individuals. J. Econ. Theory **66**(2), 615–625 (1995)

[51] S.J. Brams, M. Kilgour, Kingmakers and leaders in coalition formation. Soc. Choice Welf. **41**(1), 1–18 (2013)

[52] S.J. Brams, M. Kilgour, National security games. Synthese **76**, 185–200 (1988)

[53] V.V. Breer, D.A. Novikov, A.D. Rogatin, *Mob Control: Models of Threshold Collective Behavior* (Springer, Heidelberg, 2017) (Translated from the Russian Edition, Moscow, Lenand, 2016)

[54] P.E. Caines, Mean field games, in *Encyclopedia of Systems and Control*, ed. by T. Samad, J. Ballieul (Springer, London, 2014), p. 30–1. https://doi.org/10.1007/978-1-4471-5102-9. Reference 364780

[55] M.J. Canty, D. Rothenstein, R. Avenhaus, Timely inspection and deterrence. Eur. J. Oper. Res. **131**, 208–223 (2001)

[56] P. Cardaliaguet, F. Delarue, J.-M. Lasry, P.-L. Lions, *The Master Equation and the Convergence Problem In Mean Field Games*. https://arxiv.org/abs/1509.02505

[57] P. Cardaliaguet, J.-M. Lasry, P.-L. Lions, A. Porretta, Long time average of mean field games with a nonlocal coupling. SIAM J. Control Optim. **51**(5), 3558–3591 (2013)

[58] R. Carmona, F. Delarue, Probabilistic analysis of mean-field games. SIAM J. Control Optim. **514**, 2705–2734 (2013)

[59] R. Carmona, F. Delarue, The master equation for large population equilibriums, in *Stochastic Analysis and Applications*, vol. 100, Springer Proceedings in Mathematics and Statistics (Springer, Cham, 2014), pp. 77–128

[60] R. Carmona, F. Delarue, *Probabilistic Theory of Mean Field Games with Applications*, vol. I, II; *Probability Theory and Stochastic Modelling*, vol. 83, 84 (Springer, Berlin, 2018)

[61] R. Carmona, D. Lacker, A probabilistic weak formulation of mean field games and applications. Ann. Appl. Probab. **25**(3), 1189–1231 (2015)

[62] C. Castillo-Chavez, B. Song, Models for the transmission dynamics of fanatic behaviors, in *Bioterrorism: Mathematical Modelling Applications in Homeland Security, vol. 28, SIAM Frontiers in Applied Mathematucs*, ed. by H.T. Banks, C. Castillo-Chavez (SIAM, Philadelphia, 2003), pp. 155–172

[63] A. Cecchin, V.N. Kolokoltsov, Evolutionary game of coalition building under external pressure. Ann. ISDG **15**, 71–106 (2017). arXiv:1705.08160

[64] K. Chatterjee, B. Dutta, D. Ray, K. Sengupta, A noncooperative theory of coalition bargaining. Rev. Econ. Stud. **60**, 463–477 (1993)

[65] A. Clauset, M. Young, K.S. Gleditsch, On the frequency of severe terrorist events. J. Confl. Resolut. **51**, 58–87 (2007). arXiv:physics/0606007v3

[66] S.H. Clearwater (ed.), *Market-Based Control: A Paradigm for Distributed Resource Allocation* (World Scientific, Singapore, 1995)

[67] D. Condorelli, Market and non-market mechanisms for the optimal allocation of scarce resources. Games Econ. Behav. **82**, 582–591 (2013)

[68] V. Corradi, R. Sarin, Continuous approximations of stochastic evolutionary game dynamics. J. Econ. Theory **94**(2), 163–191 (2000)

[69] St Dereich, P. Mörters, Random networks with sublinear preferential attachent: the giant component. Ann. Prob. **41**(1), 329–384 (2013)

[70] Y. Deutsch, B. Golany, N. Goldberg, U.G. Rothblum, Inspecion games with local and global allocation bounds. Naval. Res. Logist. **60**, 125–140 (2013)

[71] H. Diamond, Minimax policies for unobservable inspections. Math. Oper. Res. **7**(1), 139–153 (1982)

[72] B. Djehiche, H. Tembine, R. Tempone, A stochastic maximum principle for risk-sensitive mean-field type control. IEEE Trans. Automat. Control **60**(10), 26402649 (2015)

[73] B. Djehiche, A. Tcheukam, H. Tembine, A mean-field game of evacuation in mutlilevel building. IEEE Trans. Automat. Control **62**(10), 51545169 (2017)

[74] U. Dobramysl, M. Mobilia, M. Pleimling, U. Täuber, Stochastic population dynamics in spatially extended predator-prey systems. J. Phys. A **51**(6), 063001 (2018)

[75] R. Dorfman, P. Samuelson, R. Solow, *Linear Programming and Economic Analysis* (McGraw-Hill, New York, 1958)

[76] M. Dresher, A sampling inspection problem in arms control agreements: a game theoretical analysis. Memorandum No. RM-2872-ARPA (Rand Corporation) (Santa Monica, 1962)

[77] D. Duffie, S. Malamud, G. Manso, Information percolation with equilibrium search dynamics. Econometrica **77**, 1513–1574 (2009)

[78] N. El Karoui, I. Karatzas, Dynamic allocation problems in continuous time. Ann. Appl. Probab. **4**(2), 255286 (1994)

[79] G. Ellison, Learning, local interaction, and coordination. Econometrica **61**(5), 1047–1071 (1993)

[80] W. Enders, T. Sandler, *The Political Economy of Terrorism* (Cambridge University Press, Cambridge, 2012)

[81] J.R. Faria, D. Arce, A vintage model of terrorist organizations. J. Confl. Resolut. **56**(4), 629–650 (2012)

[82] T.S. Ferguson, C. Melolidakis, On the inspection game. Naval. Res. Logist. **45**, 327–334 (1998)

[83] H. Fielding, *The History of the Life of the Late Mr Jonathan Wilde the Great* (H. Hamilton, 1947)

[84] A.F. Filippov, Differential equations with discontinuous right-hand side. Mat. Sb. **51(93)**(1), 99–128 (1960)

[85] A.F. Filippov, *Differential Equations with Discontinuous Righthand Sides* (Kluwer Academic, Dordrecht, 1988)

[86] W.H. Fleming, H.M. Soner, *Controlled Markov Processes and Viscosity Solutions*, 2nd edn. (Springer, Berlin, 2006)

[87] D. Foster, P. Young, Stochastic evolutionary game dynamics. Theoret. Popul. Biol. **38**(2), 219–232 (1990)

[88] M. Freidlin, A. Wentzell, *Random Perturbations of Dynamic Systems* (Springer, Berlin, 1984)

[89] A.Y. Garnaev, A remark on the customs and smuggler game. Naval. Res. Logist. **41**, 287–293 (1994)

[90] N. Gast, B. Gaujal, J.-Y. Le Boudec, Mean field for markov decision processes: from discrete to continuous optimization. IEEE Trans. Automat. Control **57**(9), 2266–2280 (2012)

[91] N. Gast, B. Van Houdt, A refined mean field approximation. Proc. ACM Meas. Anal. Comput. Syst. **1(28)**, (2017). https://doi.org/10.1145/3152542. https://hal.inria.fr/hal-01622054

[92] N. Gast, L. Bortolussi, M. Tribastone, Size expansions of mean field approximation: transient and steady-state analysis. Performance Evaluation (2018). To appear

[93] J.C. Gittins, *Multi-Armed Bandit and Allocation Indices* (Wiley, New York, 1989)

[94] F. Giovannoni, D.J. Seidmann, Corruption and power in democracies. Soc. Choice. Welf. **42**, 707–734 (2014)

[95] D.A. Gomes, J. Mohr, R. Souza, Discrete time, finite state space mean field games. J. Math. Pures Appl. (9) **93**(3), 308–328 (2010)

[96] D.A. Gomes, J. Mohr, R.R. Souza, Continuous time finite state space mean field games. Appl. Math. Optim. **68**(1), 99–143 (2013)

[97] D.A. Gomes, S. Patrizi, V. Voskanyan, On the existence of classical solutions for stationary extended mean field games. Nonlinear Anal. **99**, 49–79 (2014)

[98] D. Gomes, R.M. Velho, M.-T. Wolfram, Socio-economic applications of finite state mean field games. Philos. Trans. R. Soc. Lond. Ser. A Math. Phys. Eng. Sci. **372**(2028), 20130405 (2014)

[99] D.A. Gomes, J. Saude, Mean field games models - a brief survey. Dyn. Games Appl. **4**(2), 110–154 (2014)

[100] A. Gorban, M. Shahzad, The Michaellis-Menten-Stueckelberg theorem. Entropy **13**(5), 966–1019 (2011)

[101] A. Gorban, V. Kolokoltsov, Generalized mass action law and thermodynamics for generalized nonlinear Markov processes. The Mathematical Modelling of Natural Phenomena (MMNP) (2015). To appear

[102] O.I. Gorbaneva, G.A. Ugolnitskii, A.B. Usov, *Modeling Corruption in Hierarchical Systems of Control (in Russian)*, Monograph (South Federal University, Rostov-na-Donu, 2014)

[103] O.I. Gorbaneva, G.A. Ugolnitskii, A.B. Usov, Models of corruption in hierarchical systems of control (in Russian). Control Sci. **1**, 2–10 (2014)

[104] A. Goyal, J.B. Shukla, A.K. Misra, A. Shukla, Modeling the role of government efforts in controlling extremism in a society. Math. Meth. Appl. Sci. **38**, 4300–4316 (2015)

[105] O. Guéant, From infinty to one: The reduction of some mean feld hames to a global control problem. https://arxiv.org/abs/1110.3441

[106] E. Gubar, S. Kumacheva, E. Zhitkova, Z. Kurnosykh, T. Skovorodina, Modelling of information spreading in the population of taxpayers: evolutionary approach. Contrib. Game Theory Manag. **10**, 100–128 (2017)

[107] P.J. Graber, A. Bensoussan, Existence and uniqueness of solutions for Bertrand and Cournot mean field games. Appl. Math. Optim. **77**(1), 4771 (2018)

[108] O. Guéant, Existence and uniqueness result for mean field games with congestion effect on graphs. Appl. Math Optim **72**, 291–303 (2015)

[109] O. Guéant, J.-M. Lasry, P.-L. Lions, Mean field games and applications, in *Paris-Princeton Lectures on Mathematical Finance 2010*, Lecture Notes in Mathematics (Springer, Berlin, 2003), pp. 205–266

[110] R. Gunther, L. Levitin, B. Schapiro, P. Wagner, Zipf's law and the effect of ranking on probability distributions. Int. J. Theor. Phys. **35**(2), 395–417 (1996)

[111] J. Harsanyi, R. Selten, *A General Theory of Equilibrium Selection in Games* (MIT Press, Cambridge, 1988)

[112] O. Hernandez-Lerma, J.B. Lasserre, *Discrete-Time Markov Control Processes* (Springer, New York, 1996)

[113] O. Hernandez-Lerma, *Lectures on Continuous-time Markov Control Processes*, Aportaciones Matematicas: Textos, 3 (Sociedad Matematica Mexicana, Mexico, 1994)

[114] H.W. Hethcote, The mathematics of infectious diseases. SIAM Rev. **42**(4), 599653 (2000)

[115] J. Hofbauer, K. Sigmund, *Evolutionary Games and Population Dynamics* (Cambridge University Press, Cambridge, 1998)

[116] R. Hohzaki, An inspection game with multiple inspectees. Eur. J. Oper. Res. **178**, 894–906 (2007)

[117] E. Höpfinger, A game-theoretic analysis of an inspection problem. Technical Report, C-Notiz No. 53 (preprint), (University of Karlsruhe, 1971)

[118] M. Huang, R. Malhamé, P. Caines, Large population stochastic dynamic games: closed-loop Mckean-Vlasov systems and the Nash certainty equivalence principle. Commun. Inf. Syst. **6**, 221–252 (2006)

[119] M. Huang, P. Caines, R. Malhamé, Large-population cost-coupled LQG problems with nonuniform agents: individual-mass behavior and decentralized ϵ-Nash equilibria. IEEE Trans. Automat. Control **52**(9), 1560–1571 (2007)

[120] M. Huang, Large-population LQG games involving a major player: the Nash certainty equiv-
       alence principle. SIAM J. Control Optim. **48**, 3318–3353 (2010)
[121] L. Hurwicz, But who will guard the guardians? Prize Lecture 2007. www.nobelprize.org
[122] A.K. Jain, Corruption: a review. J. Econ. Surv. **15**(1), 71–121 (2001)
[123] B. Jovanovic, R.W. Rosental, Anonymous sequential games. J. Math. Econ. **17**, 77–87 (1988)
[124] M. Kandori, G.J. Mailath, R. Rob, Learning, mutation, and long run equilibria in games.
       Econometrica **61**(1), 29–56 (1993)
[125] S. Katsikas, V. Kolokoltsov, W. Yang, Evolutionary inspection and corruption game. Games
       7(4), 31 (open access) (2016)
[126] S. Katsikas, V. Kolokoltsov, Evolutionary, Mean-Field and Pressure-Resistance Game Mod-
       elling of Networks Security. Submitted
[127] A.N. Kochubei, Yu. Kondratiev, Fractional kinetic hierarchies and intermittency. Kinet. Relat.
       Models **10**(3), 725740 (2017)
[128] I. Kolesin, O.A. Malafeyev, M. Andreeva, G. Ivanukovich, Corruption: Taking into account
       the psychological mimicry of officials, 2017, in *AIP Conference Proceedings AIP Conference
       Proceedings 1863* (2017), p. 170014. https://doi.org/10.1063/1.4992359
[129] V.N. Kolokoltsov, Hydrodynamic limit of coagulation-fragmentation type models of k-nary
       interacting particles. J. Stat. Phys. **115**(5/6), 1621–1653 (2004)
[130] V. Kolokoltsov, Measure-valued limits of interacting particle systems with k-nary interaction
       II. Finite-Dimens. Limits Stoch. Stoch. Rep. **76**(1), 45–58 (2004)
[131] V.N. Kolokoltsov, Kinetic equations for the pure jump models of k-nary interacting particle
       systems. Markov Process. Relat. Fields **12**, 95–138 (2006)
[132] V. Kolokoltsov, Nonlinear Markov semigroups and interacting Lévy type processes. J. Stat.
       Phys. **126**(3), 585–642 (2007)
[133] V.N. Kolokoltsov, Generalized Continuous-Time Random Walks (CTRW), Subordina-
       tion by Hitting Times and Fractional Dynamics. Probab. Theory Appl. **53**(4) (2009).
       arXiv:0706.1928v1 [math.PR] 2007
[134] V.N. Kolokoltsov, The central limit theorem for the Smoluchovski coagulation model.
       Prob. Theory Relat. Fields **146**(1), 87 (2010). https://doi.org/10.1007/s00440-008-0186-
       2. arXiv:0708.0329v1 [math.PR] 2007
[135] V. N. Kolokoltsov. Noninear Markov games. Talk given at Adversarial and
       Stochastic Elements in Autonomous Systems Workshop, 22-24 March 2009, at
       FDIC, Arlington, VA 22226. Semantic Scholar. https://pdfs.semanticscholar.org/3eab/
       2cf5bcb98548bd5a862932370a3bfc792fe7.pdf
[136] V.N. Kolokoltsov, *Nonlinear Markov Processes and Kinetic Equations*, vol. 182, Cambridge
       Tracks in Mathematics (Cambridge University Press, Cambridge, 2010)
[137] V.N. Kolokoltsov, *Markov Processes, Semigroups and Generators*, vol. 38, DeGruyter Stud-
       ies in Mathematics (DeGruyter, 2011)
[138] V.N. Kolokoltsov, Nonlinear Markov games on a finite state space (mean-field and binary
       interactions). Int. J. Stat. Probab. **1**(1), 77–91 (2012). http://www.ccsenet.org/journal/index.
       php/ijsp/article/view/16682
[139] V.N. Kolokoltsov, On fully mixed and multidimensional extensions of the Caputo and
       Riemann-Liouville derivatives, related Markov processes and fractional differential equa-
       tions. Fract. Calc. Appl. Anal. **18**(4), 1039–1073 (2015)
[140] V.N. Kolokoltsov, The evolutionary game of pressure (or interference), resistance and collab-
       oration (2014). MOR (Mathematics of Operations Research) **42**(4), 915944 (2017). http://
       arxiv.org/abs/1412.1269
[141] V.N. Kolokoltsov, Chronological operator-valued Feynman-Kac formulae for generalized
       fractional evolutions. arXiv:1705.08157
[142] V. Kolokoltsov, *Differential Equations on Measures and Functional Spaces* (Birkhauser,
       2019). To appear
[143] V.N. Kolokoltsov, A. Bensoussan, Mean-field-game model of botnet defence in cyber-
       security. AMO (Applied Mathematics and Optimization) **74**(3), 669–692 (2015). http://
       arxiv.org/abs/1511.06642

[144] V.N. Kolokoltsov, O.A. Malafeyev, *Understanding Game Theory* (World Scientific, Singapore, 2010)

[145] V.N. Kolokoltsov, O.A. Malafeyev, Mean field game model of corruption (2015). Dynamics Games and Applications **7**(1), 34–47 (2017). Open Access

[146] V. Kolokoltsov, O. Malafeyev, Corruption and botnet defense: a mean field game approach. Int. J. Game Theory. https://doi.org/10.1007/s00182-018-0614-1. http://link.springer.com/article/10.1007/s00182-018-0614-1

[147] V.N. Kolokoltsov, V.P. Maslov, *Idempotent Analysis an its Applications* (Kluwer Academic, Dordrecht, 1987)

[148] V.N. Kolokoltsov, H. Passi, W. Yang, Inspection and crime prevention: an evolutionary perspective (2013). http://arxiv.org/abs/1306.4219

[149] V. Kolokoltsov, M. Troeva, W. Yang, On the rate of convergence for the mean-field approximation of controlled diffusions with large number of players. Dyn. Games Appl. **4**(2), 208–230 (2014)

[150] V. Kolokoltsov, M. Veretennikova, Fractional Hamilton Jacobi Bellman equations for scaled limits of controlled continuous time random walks. Commun. Appl. Ind. Math. **6**(1), e-484 (2014). http://caim.simai.eu/index.php/caim/article/view/484/PDF

[151] V.N. Kolokoltsov, M. Veretennikova, The fractional Hamilton-Jacobi-Bellman equation. J. Appl. Nonlinear Dyn. **6**(1), 4556 (2017)

[152] V. Kolokoltsov, W. Yang, The turnpike theorems for Markov games. Dyn. Games Appl. **2**(3), 294–312 (2012)

[153] V. Kolokoltsov, W. Yang, Existence of solutions to path-dependent kinetic equations and related forward - backward systems. Open J. Optim. **2**(2), 39–44 (2013). http://www.scirp.org/journal/ojop/

[154] N.A. Korgin, V.O. Korepanov, An efficient solution of the resource allotment problem with the GrovesLedyard mechanism under transferable utility Automation and remote control. **77**(5), 914–942 (2016)

[155] P.L. Krapivsky, S. Redner, Organization of growing random networks. Phys. Rev. E **63**, 066123 (2001)

[156] G.E. Kreindler, H.P. Young, Fast convergence in evolutionary equilibrium selection. Games Econ. Behav. **80**, 39–67 (2013)

[157] A. Lambert-Mogiliansky, M. Majumdar, R. Radner, Strategic analysis of petty corruption with an intermediary. Rev. Econ. Des. **13**(1–2), 45–57 (2009)

[158] A. Lambert-Mogiliansky, M. Majumdar, R. Radner, Petty corruption: a game-theoretic approach. J. Econ. Theory **4**, 273–297 (2008)

[159] J.-M. Lasry, P.-L. Lions, Jeux à champ moyen, I. Le cas stationnaire. Comptes Rendus Mathematique Acad. Sci. Paris **343**(9), 619–625 (2006)

[160] J.-M. Lasry, P.-L. Lion, Mean-field games with a major player. Jeux champ moyen avec agent dominant. Comptes Rendus Mathematique Acad. Sci. Paris **356**(8), 886–890 (2018)

[161] J.-Y. Le Boudec, The stationary behaviour of fluid limits of reversible processes is concentrated on stationary points. Netw. Heterog. Media **8**(2), 529–540 (2013)

[162] J.-H. Lee, K. Sigmund, U. Dieckmann, Y. Iwasa, Games of corruption: How to suppress illegal logging. J. Theor. Biol. **367**, 1–13 (2015)

[163] M.I. Levin, M.L. Tsirik, Mathematical modeling of corruption (in Russian). Ekon. i Matem. Metody **34**(4), 34–55 (1998)

[164] Zh. Li, Q. Liao, A. Striegel, Botnet economics: uncertainty matters. http://weis2008.econinfosec.org/papers/Liao.pdf

[165] J. Liu, Y. Tang, Z.R. Yang, The spread of disease with birth and death on networks. J. Stat. Mech.: Theory Exp. (2004). arXiv:q-bio/0402042v3

[166] D. López-Pintado, Contagion and coordination in random networks. Int. J. Game Theory **34**(3), 371–381 (2006)

[167] K.-W. Lye, J.M. Wing, Game strategies in network security. Int. J. Inf. Secur. **4**, 71–86 (2005)

[168] J.P. Lynch, K.H. Law, Market-based control of linear structural systems. Earthq. Eng. Struct. Dyn. **31**, 1855–1877 (2002)

[169] O.A. Malafeyev, Controlled conflict systems (In Russian) (St. Petersburg State University Publication, 2000). ISBN 5-288-01769-7

[170] O.A. Malafeyev (ed.) Introduction to modeling corruption systems and processes (in Russian), vol. 1 and 2 (Stavropol, 2016)

[171] O.A. Malafeyev, N.D. Redinskikh, G.V. Alferov, Electric circuits analogies in economics modeling: Corruption networks. in *Proceedings of ICEE-2014 (2nd International Conference on Emission Electronics)* (IEEE). https://doi.org/10.1109/Emission.2014.6893965

[172] O.A. Malafeyev, S.A. Nemnyugin, E.P. Kolpak, A. Awasthi, Corruption dynamics model, in *AIP Conference Proceedings 1863* (2017), p. 170013. https://doi.org/10.1063/1.4992358

[173] O.A. Malafeyev, D. Saifullina, G. Ivaniukovich, V. Marakhov, I. Zaytseva, The model of multi-agent interaction in a transportation problem with a corruption component, in *AIP Conference Proceedings 1863*, 2017, p. 170015. https://doi.org/10.1063/1.4992360

[174] M.A. Maschler, A price leadership method for solving the inspector's non-constant-sum game. Naval. Res. Logist. Q. **13**, 11–33 (1966)

[175] V.V. Mazalov, *Mathematical Game Theory and Applications* (Wiley, New York, 2014)

[176] V.V. Mazalov, A.N. Rettieva, Fish war and cooperation maintenance. Ecol. Model. **221**, 1545–1553 (2010)

[177] M.M. Meerschaert, A. Sikorskii, *Stochastic Models for Fractional Calculus*, vol. 43, De Gruyter Studies in Mathematics (NY, 2012)

[178] B.B. De Mesquita, *The Predictioneer's Game* (Random House, 2010)

[179] M. Mobilia, I.T. Georgiev, U.C. Täuber, C. Uwe, Phase transitions and spatio-temporal fluctuations in stochastic lattice Lotka-Volterra models. J. Stat. Phys. **128**(1–2), 447–483 (2007)

[180] E.G. Neverova, O.A. Malafeyev, A model of interaction between anticorruption authority and corruption groups, *AIP Conference Proceedings*, vol. 1648 (2015), p. 450012. https://doi.org/10.1063/1.4912671

[181] F. Ngendafuriyo, G. Zaccour, Fighting corruption: to precommit or not? Econ. Lett. **120**, 149–154 (2013)

[182] P.V. Nikolaev, Corruption suppression models: the role of inspectors' moral level. Conmput. Math. Model. **25**(1), 87–102 (2014)

[183] D.A. Novikov, *Project Management: Organizational Mechanisms (In Russian)* (PMSOFT, Moscow, 2007)

[184] J. Norris, Cluster coagulation. Commun. Math. Phys. **209**, 407–435 (2000)

[185] J. Norris. Markov Chains. Cambridge University Press

[186] M. Nourian, P.E. Caines, ϵ-Nash mean field game theory for nonlinear stochastic dynamical systems with major and minor agents. SIAM J. Control Optim. **51**(4), 3302–3331 (2013)

[187] M. Nourian, P.E. Caines, R.P. Malhamé, Mean field anlysis of controlled Cucker-Smale type flocking: Linear anaysis and perturbation equations, in *Proceedings of hte 18th IFAC World Congress, milan 2011*, ed. by S. Bittani (Curran Associates, 2011), pp. 4471–4476

[188] C. Nowzari, V. Preciado, G.J. Pappas, Optimal resource allocation for control of networked epidemic models. IEEE Trans. Control Netw. Syst. **4**(2), 159–169 (2017)

[189] M. Nutz, A mean field game of optimal stopping. SIAM J. Control Optim. **56**(2), 12061221 (2018)

[190] M.R. DOrsogna, M. Perc, Statistical physics of crime: A review. Phys. Rev. **12**, 1–21 (2015)

[191] M. Perc, J.J. Jordan, D.G. Rand, Z. Wang, S. Boccaletti, A. Szolnoki, Statistical physics of human cooperation. Phys. Rep. **687**, 1–51 (2017)

[192] I.G. Petrovski, *Ordinary Differential Equations* (Prentice-Hall Inc., N.J. 1966). Translated from the Russian and edited by Richard A. Silverman

[193] I.G. Petrovsky, *Lectures on Partial Differential Equations* (Dover Publications Inc., New York, 1991). Translated from the Russian by A. Shenitzer

[194] L.A. Petrosyan, N.A. Zenkevich, *Game Theory*, 2nd edn. (World Scietific, 2016)

[195] Y.A. Pichugin, O.A. Malafeyev, Statistical estimation of corruption indicators in the firm. Appl. Math. Sci. **10**(42), 2065–2073 (2016)

[196] D.O. Pushkin, H. Aref, Bank mergers as scale-free coagulation. Phys. A **336**, 571–584 (2004)

[197] L. Rass, J. Radcliffe, *Spatial Deterministic Epidemics*, vol. 102. Mathematical Surveys and Monographs (AMS, 2003)

[198] W. Ren, C. Yongcan, *Distributed Coordination of Multi-agent Networks* (Springer, Berlin, 2011)

[199] L.F. Richardson, Variation of the frequency of fatal quarrels with magnitude. J. Am. Stat. Assoc. **43**, 523 (1948)

[200] S. Rose-Ackerman, *Corruption: A Study in Piblic Economy* (Academic, N.Y., 1978)

[201] S. Rose-Ackerman, *Corruption and Government: Cuases, Consequences and Reforms* (Cambridge University Press, Cambridge, 1999)

[202] B.P. Rosendorff, T. Sandler, Too much of a good thing?: the proactive response Dilemma. J. Confl. Resolut. **48**, 657–671 (2005)

[203] Sh. Ross, *Introduction to Stochastic Dynamic Programming* (Wiley, New York, 1983)

[204] G. Roth, W. Sandholm, Stochastic approximations with constant step size and differential inclusions. SIAM J. Control Optim. **51**(1), 525–555 (2013)

[205] D. Rothenstein, S. Zamir, Imperfect inspection games over time. Ann. Oper. Res. **109**, 175–192 (2002)

[206] A. Saichev, Ya. Malvergne, D. Sornette, *Theory of Zipf's Law and Beyond*, vol. 632, Lecture Notes in Economics and Mathematicl Systems (Springer, Berlin, 2010)

[207] M.A. Sakaguchi, A sequential allocation game for targets with varying values. J. Oper. Res. Soc. Jpn. **20**, 182–193 (1977)

[208] F. Salten, *Bambi, A Life in the Woods* (1928). English Translation Simon and Schuster

[209] W. Sandholm, Almost global convergence to p-dominant equilibrium. Int. J. Game Theory **30**(1), 107–116 (2001)

[210] W. Sandholm, Stochastic imitative game dynamics with committed agents. J. Econ. Theory **147**(5), 2056–2071 (2012)

[211] T. Sandler, D. Arce, Terrorism and game theory. Simul. Gaming **34**(3), 319–337 (2003)

[212] T. Sandler, H.E. Lapan, The calculus of dissent: an analysis of terrorists' choice of targets. Synthese **76**(2), 245–261 (1988)

[213] F.J. Santonja, A.C. Tarazona, R.J. Villanueva, A mathematical model of the pressure of an extreme ideology on a society. Comput. Math. Appl. **56**, 836–846 (2008)

[214] A. Saperstein, Mathematical modeling of the interaction between terrorism and counter-terrorism and its policy implications. Complexity **14**(1), 4549 (2008)

[215] Y. Shoham, K. Leyton-Brown, *Multiagent Systems: Algorithmic, Game-theoretic and Logical Foundations* (Cambridge University Press, Cambridge, 2008)

[216] M.V. Simkin, V.P. Roychowdhury, Re-inventing Willis. Phys. Rep. **502**, 1–35 (2011)

[217] R. Starkermann, Unity is strength or corruption! (a mathematical model). Cybern. Syst. Int. J. **20**(2), 153–163 (1989)

[218] A.I. Subbotin, *Generalized Solutions of First Order of PDEs: The Dynamical Optimization Perspectives* (Birkhauser, Boton, 1995)

[219] A.M. Tarasiev, A.A. Usova, YuV Schmotina, Calculation of predicted trajectories of the economic development under structural changes (In Russian). Math. Theory Games Appl. (Matematicheskaya teoria ugr i priloshenia) **8**(3), 34–66 (2016)

[220] H. Tembine, Q. Zhu, T. Basar, Risk-sensitive mean-field games. IEEE Trans. Automat. Control **59**(4), 835–850 (2014)

[221] M.U. Thomas, Y. Nisgav, An infiltration game with time dependent payoff. Naval. Res. Logist. Q. **23**, 297–302 (1976)

[222] V.V. Uchaikin, *Fractional Derivatives for Physicists and Engineers* (Springer, Berlin, 2012)

[223] F. Udwadia, G. Leitmann, L. Lambertini, A dynamical model of terrorism. Discret. Dyn. Nat. Soc. (2006). Art. ID 85653

[224] I.A. Ushakov, *Optimal Resource Allocation with Practical Statistical Applications and Theory* (Wiley, NJ, 2013)

[225] A.A. Vasin, *Noncooperative Games in Nature and Society* (MAKS Press, Moscow, 2005). (in Russian)

[226] A.A. Vasin, P.A. Kartunova, A.S. Urazov, Models of organization of state inspection and anticorruption measures. Matem. Model. **22**(4), 67–89 (2010)

[227] B. Von Stengel, Recursive inspection games. Math. Oper. Res. **41**(3), 935952 (2016)

[228] B-Ch. Wang, J.-F. Zhang, Distributed output feedback control of Markov jump multi-agent systems. Autom. J. IFAC **49**(5), 1397–1402 (2013)

[229] B.J. West, *Fractional Calculus View of Complexity. Tomorrow's Science* (CRC Press, Boca Raton, 2016)

[230] P. Wiecek, Total Reward Semi-Markov Mean-Field Games with Complementarity Properties (2015). Submitted for publication

[231] S. Wrzaczek, E. Kaplan, J.P. Caulkins, A. Seidl ans G. Feichtinger. Differential terror queue games. Dyn. Games Appl. **7**, 578–593 (2017)

[232] G. Yaari, A. Nowak, K. Rakocy, S. Solomon, Microscopic study reveals the singular origins of growth. Eur. Phys. J. B **62**, 505–513 (2008). https://doi.org/10.1140/epjb/e2008-00189-6

[233] H. Yin. P.G. Mehta, S.P. Meyn, U.V. Shanbhag. Synchronisation of coupled oscillators is a game. IEEE Trans. Autom. Control **57**(4), 920–935 (2012)

[234] H.P. Young, The evolution of conventions. Econometrica **61**(1), 57–84 (1993)

[235] A.J. Zaslavski, *Turnpike Properties in the Calculus of Variations and Optimal Control* (Springer, New York, 2006)

[236] J. Zhang, Approximately optimal computing budget allocation for selection of the best and worst designs. IEEE Trans. Autom. Control **62**(7), 3249–3261 (2017)

Index

© Springer Nature Switzerland AG 2019
V. N. Kolokoltsov and O. A. Malafeyev, *Many Agent Games in Socio-economic Systems: Corruption, Inspection, Coalition Building, Network Growth, Security*, Springer Series in Operations Research and Financial Engineering, https://doi.org/10.1007/978-3-030-12371-0

Printed in the United States
By Bookmasters

Printed in the United States
By Bookmasters